T0091493

OUR ANIMAL CONNECTION
Second Edition

OUR ANIMAL CONNECTION

What Sapiens Can Learn from Other Species

Second Edition

Michael Hehenberger

Zhi Xia

Foreword by Huanming Yang

JENNY STANFORD
PUBLISHING

Published by
Jenny Stanford Publishing Pte. Ltd.
Level 34, Centennial Tower
3 Temasek Avenue
Singapore 039190

Email: editorial@jennystanford.com
Web: www.jennystanford.com

British Library Cataloguing-in-Publication Data
A catalogue record for this book is available from the British Library.

Our Animal Connection: What Sapiens Can Learn from Other Species (Second Edition)
Copyright © 2021 by Jenny Stanford Publishing Pte. Ltd.

All rights reserved. This book, or parts thereof, may not be reproduced in any form or by any means, electronic or mechanical, including photocopying, recording or any information storage and retrieval system now known or to be invented, without written permission from the publisher.

For photocopying of material in this volume, please pay a copying fee through the Copyright Clearance Center, Inc., 222 Rosewood Drive, Danvers, MA 01923, USA. In this case permission to photocopy is not required from the publisher.

ISBN 978-981-4877-50-3 (Hardcover)
ISBN 978-1-003-13072-7 (eBook)

We dedicate this book to the planet Earth,
hoping that sapiens will learn to respect other living species and
understand that the future of our planet is more important than
disagreements among nations.

Contents

Foreword: Learn from Others

We, homo sapiens, are life, a part of this beautiful world.

We, together with all other animals and living species, are the creation of Nature. But we have also transcended her in a sense. From a product of circumstances, we have risen to responsibility: the responsibility for ourselves, for all other species, for the whole environment where we live and on which we depend.

This book is a thanksgiving to other life types or species. We have learned so much from them since the very beginning, as a Chinese proverb goes: To learn from the laws of Nature Teacher (师法自然).

There is great diversity among Animals. It is estimated that there are about 8.7 million species, all unique and different from each other, morphologically or anatomically. However, Animals also have a lot in common, starting with shared biology based on cell structure and function. Some animals have limbs, wings, hearts, lungs, gills, bones, and others have not. Their body sizes can vary from tiny unicellular mycoplasma to the giant blue whale. Animals can be found everywhere on our planet, they have conquered every ecosystem, showing an amazing ability to adapt to even the most extreme habitat.

Along with plants and microorganisms, Animals have survived through all our planet's eras since the origin of life on Earth. Earth has been an ever-changing place. Some species adapted to change and survived, others became extinct. During millions of years, Animals evolved and developed their preferences and capabilities, often in a symbiotic relationship with each other and the world of plants.

What differentiates humans is our powerful ability to communicate by means of a language, our ability to do complex reasoning and solve difficult problems. Based on our highly developed brain, we are able to learn from others and to build on existing knowledge as we move forward and penetrate ever deeper into the unknown. Our ability to adapt to every climate by wearing all kinds of clothes, by building elaborate houses, and by using energy sources to make us feel warm and comfortable when it's cold, or nice and cool when it's hot and humid outside. In addition to our language skills, our ability to use representational symbols to convey information and our mastery of abstract topics such as mathematics and philosophy are other important differentiators.

However, are our cognitive abilities qualitatively different from those of other animals, or are they just extensions of what we can observe in animals? We are not the only species to use tools. Dolphins, crows, and even sea otters, make tools the way humans do. We are not the only species able to change the environment around us to fit our needs. We know that beavers can build dams that utterly transform the way water moves through forests, and ants can build massive underground facilities.

Have humans been deceiving themselves for thousands of years that they are the only intelligent species in the animal kingdom, despite more and more evidence to the contrary? Actually, the more we learn about animals, the more we learn that animals can have superior cognitive faculties when compared to us humans. Their senses of perception often exceed ours. They may not be able to fully integrate everything they see and hear and taste and smell and touch as well as we do, but they definitely possess intelligence and they are also capable of love and sorrow. Their emotional world is still not well known, but it definitely exists.

There is no question about the importance of Animals to humans: We could not survive without them. Without the animals, our society could not have advanced to the point it has today. We hunted animals as food for millions of years, long before the formation of permanent settlements. Some of the earliest animals that we used for our food were insects, fish, wild pigs, and deer or antelope. Before the invention of agriculture or farming, we were hunting animals. Later, we raised pigs, cattle, goats and sheep and gradually decreased and eventually eliminated the need for "non-stop hunting." Animals have then been used by people to help with a variety of tasks such as the use of fur and wool to create clothing. Animals further played an important role in all sorts of heavy labor tasks: before the development of cars, horses, donkeys and oxen provided our means of transportation, carrying heavy weights materials for long distances. Another step was the domestication of the wolf to assist in hunting. Dog breeds were then taught to specialize in all kinds of useful tasks, from protection of private property to the location of hunted birds or deer in tall brush. Animals such as falcons and eagles were even trained to chase, kill and retrieve animals, helping humans to avoid difficult terrain and injury.

Animal bladders were used to create bags for drink or fill items, while hollowed-out horns from animals could be used to transmit sounds over very long distances. Bird feathers were used to balance arrows, or make clothing warmer. Apart from providing effective labor and a wide variety of tools, Animals such as dogs and cats accompany us, give us warmth and peace of mind. Dogs were domesticated to help humans with hunting, but quickly became key family members. Cats were originally domesticated to kill mice, but soon their effective companionship led humans to keep them inside their homes, just for fun and comfort. Research studies have now shown that pet owners, especially dog owners, live longer and happier lives than those without pets.

It is easy to realize now that humans and animals are created to complement each other. Every time when we look up at the sky and see birds, they teach us how to fly. Every time when watch the beautiful streamlined shape of the whale, we want to learn how to swim or build devices that take us into the depths of the ocean. Humans were inspired by animals to build airplanes, helicopters, submarines, and to develop technologies such as optical devices, Radar, and even weapons such as harpoons. We have used our Animal teachers to learn from them.

Human beings should pay tribute to animals. They are giving us strength and comfort. Their ups and downs, as well as mysterious stories, have been embedded in our life and contribute to the beauty and joy of human life. Recently, they have also started to show us how to improve our quite fragile human health.

In order to learn more about animals and to stimulate more research about potentially helpful animal qualities and capabilities, my friend Michael Hehenberger has proposed the idea to write this fantastic book. The result is a combined effort of an experienced Western scientist and his young Chinese colleague, Zhi Xia. Both share a passion for the topic and for putting their thoughts on paper. The book is not only providing a bridge between the human and animal worlds, but is also the fruit of East-West scientific collaboration. Its goal is to let us know the animals more deeply, to let us go nearer and closer to them. In its chapters, this book depicts the colorful animal world, and demonstrates their greatness and immortality.

Through a comprehensive description of a few carefully selected animals, this book allows the mystery of the animal world to slowly but clearly unfold in front of us. After reading this book, in addition to increasing his or her knowledge of animals, the reader will further receive details of recent scientific progress based on the study of animals. This progress manifests itself in new findings across disciplines such as biology, genetics and physiology. Scientists have a responsibility to behave in an ethical manner when studying animals or testing new hypotheses on animals. Written for the general public but employing scientific arguments, this book also attempts to convey to readers how scientists think, how important it is to respect facts and to always make an honest effort to apply unbiased analysis to the interpretation of facts.

The new experiences gained by studying Animals should enable humans to enhance our biological and medical knowledge, hopefully to the benefit of patients suffering from diseases. A deeper understanding of animal capabilities and their relevance for human well-being should also teach us respect for the environment and tell us to reduce the damage humans are currently inflicting on the planet Earth. And perhaps, learning more about these special Animals will also touch our readers' hearts.

In conclusion, the book tells a lot of vivid and interesting animal stories from the perspective of the animals themselves, and makes a series of comments which are related to the human life. Nature is beautiful, but behind the beauty is a world full of cruel competition. The creatures that are wandering in nature must always be vigilant. If you are more vigilant, you will have a better hope of survival. If you don't pay attention you may be moving toward your death. Animals are like this; so are humans.

Huanming Yang

About BGI and Prof. Yang

BGI was founded in 1999 with the vision of using genomics to benefit humankind and has since become the largest genomic organization in the world. With a focus on research and applications in the fields of healthcare, agriculture, conservation, and the environmental, BGI has a proven track record of innovative, high profile research. Since 1999, BGI has generated over 1000 publications, many in top-tier journals such as *Nature* and *Science*. BGI also contributes to scientific communication by publishing the international research journal *GigaScience*.

BGI's achievements have made a significant contribution to the development of genomics throughout the world. BGI's goal is to make state-of-the-art genomics (and other "omics") highly accessible to the global research community and clinical markets by integrating the industry's broadest array of leading technologies, including BGI's own BGISEQ sequencing platform, economies of scale, and expert bioinformatics resources.

BGI also offers a wide portfolio of transformative genetic testing products across major diseases, enabling medical providers and patients worldwide to realize the promise of genomics-based diagnostics and personalized healthcare.

BGI is headquartered globally in Shenzhen, China, with European headquarters in Copenhagen, Denmark; US headquarters in Cambridge, Massachusetts; Japanese headquarters in Kobe; and an Asian office in Hongkong.

Prof. Huanming Yang is a co-founder of BGI and chairman of the Board of BGI. He initiated and continues to chair the International Conference on Genomics (ICG), which has been held for 13 years in Hangzhou and Shenzhen, and is recognized and attended by leaders from across the world. The theme for the ICG meetings has been "Omics for All," a clear message to the scientific world that new knowledge leading to a deeper understanding of life must be communicated and shared.

Acknowledgments

We'd like to thank Wikimedia Foundation and the Nobel Prize Foundation for providing access to a rich reservoir of scientific facts, figures and pictures. Wikipedia is an increasingly informative and accurate source of important facts about "our animal connection." *Science* magazine and *Nature* are other sources that ever more frequently feature interesting new scientific results that confirm our thesis that sapiens can learn a lot from the study of animals. Some fun facts about popular animals were taken from popular websites cited throughout the book. We give the editors of those sites credit for presenting interesting facts about animals in a highly entertaining way.

Michael H. would further like to acknowledge many discussions with fellow hikers of Westport's Y's Men. In particular, he has enjoyed conversations with Marty Yellin, Sal Mollica, Herb Auslander, Roy McKay, Malcolm Davies, Blair McCaw, Larry Lich, Dave Bue, and Chris Lewis. They have all retained a youthful sense of curiosity about, and willingness to dig deeply into, new scientific findings. He would also like to thank his wife Ulla and his daughters Lisa and Anna for valuable discussions and suggestions. Anna even helped with proof reading despite her busy schedule. A special thanks also to his daughter Karin Hehenberger for early discussions that actually led to the idea to write *Our Animal Connection*.

When discussing the idea with Prof. Huanming Yang, "Henry" kindly agreed to lend his help and the prestige of BGI to this project, thereby strengthening its scientific underpinnings. Our decision to collaborate and to invite Zhi Xia into our project has now resulted in this book. I am deeply grateful to both my Chinese partners for working so hard, for contributing so much, and for pushing the project forward.

Chapter 1

Introduction

About 2600 years ago, the Greek storyteller Aesop started collecting "fables" where animals were used to project human qualities, and to teach moral values. Thanks to Aesop and his followers—in particular the French fabulist Jean de la Fontaine—we are convinced that owls are wise, foxes are smart, ants are diligent, and crickets happy, but lazy. We also tell fairytales to our children that teach them to be afraid of the "big bad wolf."

In this book we intend to explore another side of our human interest in animals. We will focus on benefits to human health and well-being that can possibly be derived from a deeper understanding of their evolutionary characteristics.

Considering the diversity of living organisms around us, and the role we humans are playing on top of the animal "food chain," it may be tempting to dismiss the notion that animals could teach us things we do not already know.

However, while enjoying a superior brain, humans don't necessarily excel in all other areas of physical ability: When examining how we use our five senses—vision, hearing, taste, smell, touch—we can easily find animals that outperform us in each category. Actually, we had to use our creative brain throughout human history to compensate for our shortcomings, to invent and develop ever more sophisticated devices and technologies, designed to first emulate

Our Animal Connection: What Sapiens Can Learn from Other Species (Second Edition)
Michael Hehenberger and Zhi Xia
Copyright © 2021 Jenny Stanford Publishing Pte. Ltd.
ISBN 978-981-4877-50-3 (Hardcover), 978-1-003-13072-7 (eBook)
www.jennystanford.com

and then often surpass specialized animal capabilities. For example, human visual perception can be greatly enhanced by microscopes, by binoculars and by advanced telescopes built to explore the universe. Without those devices, we could not compete with birds of prey, or even certain insects.

Although we are proud of our athletic abilities—we can run, we can jump, we can swim, we can climb mountains—our best Olympic performances are often lagging behind potential animal competitors! Our resistance to diseases and our ability to recover from injuries are other areas where our performance is not always impressive.

In this book we ask ourselves the question whether human physiology and medicine could benefit from a closer study of highly specialized and well-functioning animal capabilities. There is nothing new about a focus on biology, zoology and animal health, but what connects animals to humans may not always have been studied in sufficient detail. In particular, as we are finding surprising similarities in the genetic makeup of humans and other living organisms, there may be emerging opportunities to benefit even more from the study of species other than *homo sapiens*. Those possible benefits could be relevant for both human wellness and disease.

During millions of years, evolution has transformed many species into highly specialized and capable living organisms. We humans have only recently started evolving along with them, and today gained a dominating position. As we are trying to overcome our many weaknesses, we should accept with gratitude what evolution can teach us about "best practices."

Before introducing homo sapiens, we will review life on Earth, as it evolved on our planet for about 4 billion years. We will also cover a few examples of special adaptations made by living species to extreme conditions on the planet Earth. The sapiens' connections to other species will then be grouped into the following categories:

(i) model organisms that help us to increase our understanding of genetics, molecular biology, cellular biology, perception, neuroscience, physiology and medicine;

(ii) animal species that can teach us lessons related to our understanding and our attempts to cure human disease;

(iii) animals with advanced capabilities that could potentially extend and "improve" human performance in areas sometimes referred to as "bionic."

After that, we will review genetic engineering technologies that are already in current use or hold the promise of future transfer of capabilities from animals to humans.

Finally, a word of warning to readers based in the USA: We have chosen to use the metric system when discussing distances (measured in meters, kilometers, centimeters, etc.), volumes (liters), weights (grams, kilograms, etc.), speeds (meters per second, kilometers per hour, etc.), and temperatures (centigrades or Celsius degrees). We just felt that our international readers would not appreciate lengthy conversions into miles, yards, inches, ounces, etc., whenever a metric number is mentioned. However, a discussion of both systems is included as an appendix.

A book about our animal connection would not be complete without a chapter covering diseases that are transmitted from animals to humans. Most infectious diseases (including pandemics such as COVID-19) fall into that category.

Chapter 2

Evolution of Life on Earth

What is Life?

It took centuries of biological, chemical and geological research to get close to answering this fundamental question. Although we have learned a lot, we are still not quite sure how life started on the planet Earth. In the late Hadean period, more than 4 billion years ago, the atmosphere consisted largely of water vapor, nitrogen, and carbon dioxide, and smaller amounts of carbon monoxide, hydrogen, and sulfur compounds. There were oceans, but they were very hot, with temperatures about 100°C. The Earth was a water world with a still nonexistent continental crust, an extremely turbulent atmosphere, and a hydrosphere that was exposed to intense ultraviolet (UV) light. It was a chemical laboratory environment, almost impossible to replicate today.

What scientists seem to agree upon is that the first primitive life forms existed without access to oxygen, that there were no enzymes present to facilitate chemical reactions, and that there was no DNA available to create proteins. Organic molecules were synthesized from what we would today characterize a hostile, inorganic environment. But somehow it did happen, and there is now speculation that something similar to our definition of life started with RNA synthesis. The process further required encapsulation

Our Animal Connection: What Sapiens Can Learn from Other Species (Second Edition)
Michael Hehenberger and Zhi Xia
Copyright © 2021 Jenny Stanford Publishing Pte. Ltd.
ISBN 978-981-4877-50-3 (Hardcover), 978-1-003-13072-7 (eBook)
www.jennystanford.com

and the generation of replicates. The next step was to create a cell within a lipid wall, with DNA and the first ribosome-like mechanism to produce the 20 basic amino acids.

Before going any further, let's review a few fundamental facts about life, as shared by all species alive today. We need to understand life "from the ground up," starting with DNA, proteins, and cells.

While DNA provides the blueprint for living organisms, it's the proteins that do the heavy lifting. In the human body, there are around 25,000 genes encoded in the DNA, but the total number of proteins is much larger, perhaps 500,000, or maybe even a million.

Each DNA molecule contains a large number of nucleotides, composed of one of four nitrogen-containing nucleobases, either cytosine (C), guanine (G), adenine (A), or thymine (T); a sugar called deoxyribose; and a phosphate group. DNA stands for "deoxyribonucleic acid."

We promise not to dive much deeper into the details of molecular biology, but if the reader wants to learn more, we recommend the Nobel Foundation's website (www.nobelprize.org) as a treasure trove of firsthand discovery stories, told by the Nobel Laureates themselves. It has been a Swedish tradition to never allow Nobel Prize winners to leave Stockholm without lecturing about their respective breakthroughs, thereby stimulating generations of curious students. We will provide numerous links to this site. In addition, if the reader feels an urge to explore the intricate chemistry of atoms, molecules, and proteins, please feel free to check out a recent book by one of the authors.[1]

For our current purpose, we just need to remember that the two separate polynucleotide strands are bound together such that A is always connected with T, and C is connected with G, to form "double-stranded" DNA. Because of its unique double-helical structure (see Fig. 2.1), DNA can easily replicate and break up to initiate the transcription of "genes" into "proteins."

The DNA of an organism is divided into a "coding" part, containing the genes that can produce proteins, and a "noncoding" part, which performs less obvious tasks. The role of genes goes beyond their association with proteins. They are often referred to as "units of heredity." They typically form a region of DNA that influences a particular characteristic in an organism. The collection of all genes in a given organism is called its "genome," and its study is called "genomics."

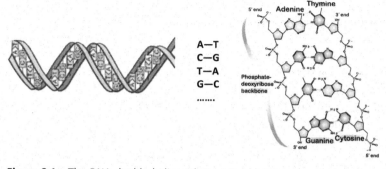

A—T
C—G
T—A
G—C
....

Figure 2.1 The DNA double helix with its A-T and C-G connections (source: Wikipedia).

In most living organisms, DNA is physically divided into "chromosomes,"[2] packaged and organized structures among which DNA is distributed.

Genetics is the study of genes, genetic variation, and heredity in living organisms. As first observed in the 1850's by Gregor Mendel, who studied flower colors of pea plants, inheritance in organisms occurs by passing discrete heritable units (genes) from parents to offspring. Mendel spent years carefully observing his peas and found that the flowers of pea plants were either purple or white—but never a shade of color in between.

Instead of peas, Thomas Hunt Morgan later studied fruit flies at his Columbia University lab in New York. Between 1910 and 1915 he discovered that each gene resides at a specific site on the chromosomes and that chromosomes are paired, one coming from the mother and the other from the father. The offspring of fruit flies (*Drosophila melanogaster*), and of other higher organisms, receives one copy of each gene from each of its two parents. Different and discrete versions of the same gene are called *alleles* and the set of alleles for a given organism is called its *genotype*. On the other hand, the openly observable features of the organism are called its *phenotype*.

In the 1920s the world of physics was revolutionized by quantum theory, by the realization that atomic energy levels are not continuous but quantized, that particles are discrete packets of energy with wavelike properties. One of the pioneers in this field was Erwin Schrödinger, and it is no surprise that he developed similar deep ideas when, after his emigration in 1940 to Dublin, Ireland, he turned his attention to biology. In his book *What Is Life*,[3] he observed that it is the difference in their genes that distinguish one animal species from another. He wrote that "genes endow organisms with their distinctive features. They code biological information in a stable form so that it can be copied and transmitted from generation to generation. The storing and passing on of biological information are carried out through the replication of chromosomes and the expression of genes."

Schrödinger's ideas helped extend biochemistry from a discipline concerned with the role of enzymes in production and utilization of energy within a cell, to a discipline concerned with copying, transmission, and editing of information.

In 1944, scientists at the Rockefeller Institute in New York discovered that genes are not proteins but that they are made of DNA.

Watson and Crick's discovery of the double helical structure of DNA then put Schrödinger's ideas into a molecular framework and confirmed that an essential role of genes is replication: at the end of their famous paper[4] they stated that "the specific pairing we have postulated immediately suggests a copying mechanism for genetic material."

When two strands of DNA unwind during replication, each parent is pairing with a complementary daughter strand. Subsequently, the daughter strand will serve as a template for the formation of still another strand, suggesting that multiple copies of DNA can be replicated as a cell is divided. By extension, the whole organism can be replicated from generation to generation. Watson and Crick further proposed that the sequence of nucleotide bases in each gene carries the code for protein synthesis.

Proteins are big molecules made up of about twenty basic building blocks, the "amino acids." Cracking the "genetic code," i.e., figuring out how instructions contained in the genes of DNA are translated into amino acids that then combine to form proteins, has been one of the major intellectual achievements of the 20th century.[5] As already stipulated by Francis Crick in his 1962 Nobel Lecture, "the genetic code describes the way in which a sequence of twenty or more things is determined by a sequence of four things of a different type."[6]

In 1956, Francis Crick also proposed the "central dogma of molecular biology," illustrated in Fig. 2.2, three years after his and James Watson's discovery[7] of the double-helical structure of DNA.

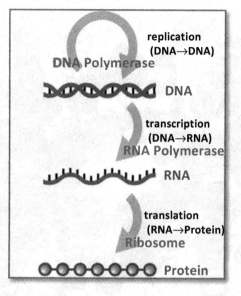

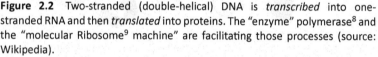

Figure 2.2 Two-stranded (double-helical) DNA is *transcribed* into one-stranded RNA and then *translated* into proteins. The "enzyme" polymerase[8] and the "molecular Ribosome[9] machine" are facilitating those processes (source: Wikipedia).

Depending on the "sequence" of the amino acids, proteins then "fold" in many different ways, resulting in often complex, and frequently spectacularly beautiful, three-dimensional structures. Those structures determine the functional characteristics of proteins, and their ability to control almost everything related to an organism's life and death. For example, Fig. 2.3 shows hemoglobin,[10] the iron-containing oxygen-transport metalloprotein which—in the blood of most complex organisms, including humans—carries oxygen from the respiratory organs (lungs or gills) to the rest of the body (i.e., the tissues). The "heme" groups contain iron and each iron atom can bind oxygen. Hemoglobin releases the oxygen to permit aerobic respiration to provide energy to power the functions of the organism in the process called "metabolism."

But we are getting ahead of ourselves—something that's hard to avoid when exploring the wonderful world of atoms, molecules, and "very big biological molecules": proteins.

To get back to our task—understand the basics of life on the planet Earth—we need to introduce the concept of "cells."

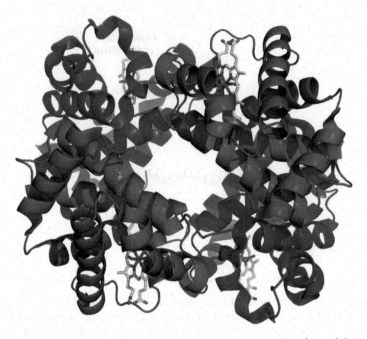

Figure 2.3 Three-dimensional hemoglobin structure with 4 (green) heme groups (source: Wikipedia).

Cells are the basic structural, functional, and biological units of all known living organisms. Living organisms can replicate, and cells are the building blocks of all organisms. Cells come in many different shapes and forms but are all constructed in a similar way: They have an outer enclosure called the membrane, a nucleus containing the necessary genetic information, and a space in between called the cytoplasm. Membranes are made up of lipids, i.e., fatty materials with very interesting properties that control what's moving in and out of the cell. The DNA is protected inside but continuously attacked, damaged, and repaired. Proteins are active everywhere and in all possible ways, including playing the roles of "enzymes" that facilitate chemical reactions and biological processes.

Organisms can be classified as unicellular (consisting of a single cell, such as bacteria) or multicellular (including plants and animals). The number and variety of cells in plants and animals varies from species to species, reaching about 40 trillion in humans.

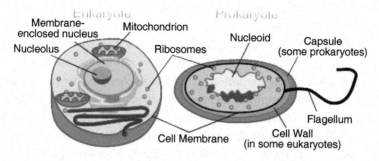

Figure 2.4   Eukaryote and prokaryote cell types (source: Wikipedia).

Prokaryotic cells (see Fig. 2.4) were the first proven form of life on Earth. They are capable of "cell signaling," and they are self-sustaining. Instead of a "cell nucleus," they contain a nuclear region called the "nucleoid." "Cell signaling" is the molecular biology term used to describe a complex system of communication that governs basic activities of cells and coordinates cell actions. Prokaryotes include two of the "domains" of life, bacteria and archaea (single-celled microorganisms, able to survive in harsh environments). The DNA of a prokaryotic cell consists of a single chromosome that is in direct contact with the cytoplasm. Prokaryotes are among the smallest of all organisms, ranging from 0.5 to 2.0 μm (micrometer, 0.000001 or 10^{-6} m) in diameter.

Also shown in Fig. 2.4 is the other major cell type, the "eukaryote."

But before turning to the eukaryotes, let us take a brief look at archaea, a very interesting form of life that has received increased attention recently.

Archaea were initially classified as bacteria, even receiving the name archaebacteria, but are now recognized for their unique properties separating them from the other two domains of life, bacteria and eukaryota.

Archaea and bacteria are generally similar in size and shape. However, archaea possess genes and several metabolic pathways that are more closely related to those of eukaryotes. In particular, archaea and eukaryotes share enzymes involved in transcription and translation. The biochemistry of archaea is further associated with lipids in their cell membranes, and with a unique ability to use a wider range of energy sources, as compared to eukaryotes. Those energy sources include organic compounds (such as sugars), ammonia, metal ions, or even hydrogen gas. Salt-tolerant archaea use sunlight as an energy source. They reproduce asexually by binary fission, fragmentation, or budding. Archaea were initially viewed as "extremophiles" living in harsh environments, such as hot springs and salt lakes, but they have since been found in a broad range of habitats, including soils, oceans, marshlands, and even the human colon. For instance, methanogens inhabit human guts, aiding digestion, and can be found in the mouth and on the skin. Archaea are particularly widespread in the oceans. As part of "plankton,"[a] the archaea may be one of the most abundant groups of organisms on the planet. Enzymes from archaea that can endure high temperatures and organic solvents are exploited in biotechnology!

In 2015, a "missing link,"[11] "Lokiarchaeota," between archaea and eukaryota was found by T. Ettema et al., who discovered this previously unknown life form near a hydrothermal (temperature 300°C) vent in the Arctic Ocean, in an underwater vent field named after the Nordic god Loki, situated between Greenland and Norway at a depth of 2352 m. According to Nordic mythos, Loki is a shape shifter, appearing in the form of a salmon, a mare, a fly, and even an elderly woman. Appropriately, the archaea named after Loki is a highly complex and confusing form of life. Analysis of Lokiarchaeota's genes and proteins have yielded the following results: Of its 5381 protein coding genes, roughly 32% do not correspond to any known protein, 26% closely resemble archaeal proteins, and 29% correspond to bacterial proteins. That suggests there has been significant inter-domain gene transfer between bacteria and archaea. Furthermore, a small but significant portion of the proteins is very similar to eukaryotic proteins, with known functions such as cell membrane deformation and cell shape formation. Most importantly, Lokiarcheota has a gene coding for *actin*, a protein that has the ability to engulf and consume another particle. Such ability could be the first step towards the *endosymbiotic* origin of mitochondria and chloroplasts (see below), a key difference between prokaryotes

---

[a]A collection of organisms including bacteria, archaea, algae, etc., that inhabit oceans, seas, or bodies of fresh water.

and eukaryotes. Endosymbiotic theory postulates the evolution of eukaryotic cells from prokaryotic organisms.

The cell membrane–related functions of Lokiarchaeum suggest that the common ancestor to the eukaryotes might be an intermediate step between the prokaryotic cells (without subcellular structures) and the eukaryotic cells.

Eukaryotic cells are about fifteen times wider than a typical prokaryote and can be as much as a thousand times greater in volume. They feature membrane-bound "organelles" (compartments) where specific metabolic activities take place. The most important organelle is the cell nucleus, which houses the cell's DNA and is the reason why "eukaryote" can be translated to "true kernel."

The "mitochondrion" deserves to be described and explained in some detail, since it generates most of the cell's supply of adenosine triphosphate (ATP), used as a source of life-sustaining chemical energy. It exists independently and carries its own (small) genome that shows substantial similarity to bacterial genomes. We will return to ATP below and tell the fascinating story how chemists and biologists developed our understanding of ATP-related chemical reactions and biological processes.

As life developed on our planet, mitochondria developed their ability to convert food and oxygen into energy. Although much more advanced and complex, eukaryotic cells could not improve on the efficiency of mitochondria, which is why they are incorporated into the cell, preserving their own genetic instructions. Plants and animals could not live without them.

Another important insight required to understand cellular function was Jacob and Monod's discovery[12] in the early 1960's of "genetic regulatory mechanisms in the synthesis of proteins." They discovered that, although every gene of the genome is present in every cell, cells are different and specialized because of "gene expression." What differentiates specialized cells in complex organisms is that certain genes are turned on and others are "turned off," leading to different and unique mixes of proteins. The unique protein mix of, say, a brain cell, enables its biological function, and so does every other specialized cell in a complex organism. Genes are switched on and off as needed to achieve optimal functioning of the cell. Those involved in the production of energy will be needed at all time; hence they will always be expressed and the proteins they encode will be part of all cellular mixes. However, genes that are not needed to perform a specialized cell's function will be shut off. Working with bacteria at the Pasteur Institute, Jacob and Monod further found that "regulatory genes" encode proteins that "regulate" the work of the "effector genes." Enzymes and ion channels are encoded by effector genes, and effector genes include control regions that are also known as "promotors." Regulatory proteins bind to the promoter of effector sites and determine whether the effector genes are switched on or off. For instance, they studied the bacterium *E. coli* and found that, depending on the environment, the gene for lactose metabolization is only turned on if lactose is present.

Again, What is Life?

What we mean by life is defined as an organism that maintains homeostasis, is composed of cells, undergoes metabolism, can grow and adapt to its environment, respond to stimuli, and reproduce. *Homeostasis* is the property of a system in which a variable (e.g., the body temperature or the pH value) is actively regulated to remain very nearly constant. The three main purposes of *metabolism* are (i) the conversion of food to energy, (ii) the conversion of food to amino acids/proteins, lipids, nucleic acids, and some carbohydrates, and (iii) the elimination of nitrogenous wastes. *Reproduction* (or procreation or breeding) is the biological process by which "offspring" are produced from their "parents." Each individual organism exists as the result of reproduction.

How did life evolve on the planet Earth?

On a "billion-year (Ga) scale" (Figs. 2.5a and 2.5b), it is clearly seen how long it took for advanced life forms to develop.

There was no life on Earth when it formed, 4.56 billion years ago. Up until some 4 billion years ago, during the Hadean geologic era, the Earth was very hot and uninviting.

During the Archean geologic era, starting about 4 billion years ago, prokaryotic cells came into existence. Actually, the first organisms may have been "chemoautotrophs": they used carbon dioxide as a carbon source and oxidized inorganic materials to extract energy. This way of creating energy was not optimal in terms of efficiency, but it was the only way possible, due to lack of oxygen in the atmosphere. Later, prokaryotes evolved glycolysis, a set of chemical reactions that freed the energy of organic molecules such as glucose and stored it in the chemical bonds of ATP. The first appearance of DNA goes back about 3.9 billion years. DNA and RNA contain the genetic information to both "run the cell" and also "replicate the cell."

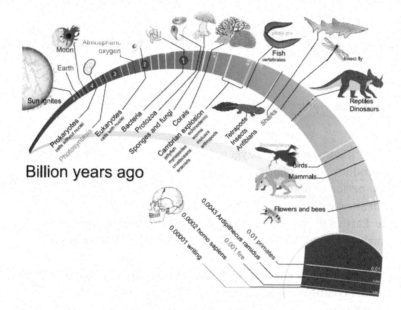

Figure 2.5a   Evolution of life on the planet Earth (source: Wikipedia).

The first prokaryotic cells had a big impact on the planet Earth because of photosynthesis.

Photosynthesis is a process used by plants and other organisms to convert light energy into chemical energy. It can later be released to fuel various activities related to "energy transformation." Chemical energy is stored in carbohydrate molecules, such as sugars, formed by combining carbon dioxide with water, as shown in Fig. 2.6.

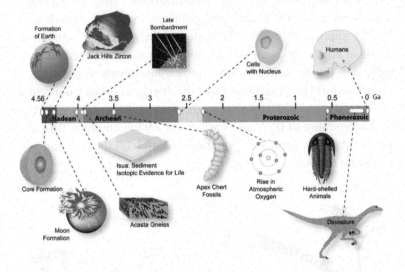

Figure 2.5b Earth's and Moon's formation and evolution of life (source: Wikipedia).

In photosynthesis, oxygen is also released as an important byproduct. Most plants, most algae, and above all *cyanobacteria*, can perform photosynthesis. By producing oxygen, the simple prokaryotic cyanobacteria have played a decisive role in converting the Earth's early oxygen-poor atmosphere, into an oxidizing one, causing the "rusting of the Earth," mostly oxidation of dissolved iron in the oceans, creating iron ore.

Photosynthesis is responsible for supplying organic molecules, and for creating and maintaining the oxygen content of the Earth's atmosphere. Life on our planet would not have evolved and would not be possible without it.

Independent of the species, the process of photosynthesis always begins when energy from light is absorbed by proteins that contain green *chlorophyll* pigments. In plants, these proteins are held inside organelles called chloroplasts. Chloroplasts are most abundant in leaf cells of plants. However, in bacteria they are embedded in the plasma membrane. Figure 2.7 shows how chlorophyll is concentrated in chloroplasts.

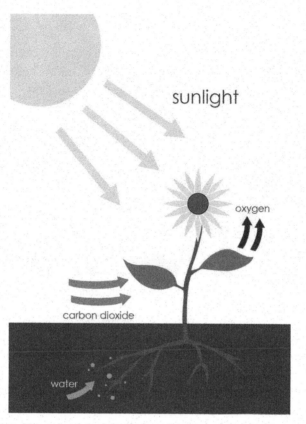

sunlight

oxygen

carbon dioxide

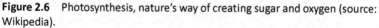

water

Figure 2.6 Photosynthesis, nature's way of creating sugar and oxygen (source: Wikipedia).

Chlorophyll absorbs light most strongly in the blue portion of the visible electromagnetic spectrum. It therefore reflects green, producing the green color of chlorophyll-containing tissues. The function of the reaction center of chlorophyll (see detailed explanation provided by 1988 Nobel Laureates J. Deisenhofer, R. Huber, and H. Michel)[13] is to absorb the sun's light energy and transfer it to other parts of the "photosystem." Eventually, some energy is used to strip electrons from water, producing oxygen gas.

In Fig. 2.5a, it is indicated how ~3.5 billion years ago the Earth's atmosphere was enriched with oxygen because of photosynthesis.

Photosynthesis then made it possible for eukaryotic cells to evolve. All organisms first developed in water. Land life is less than 500 million years old.

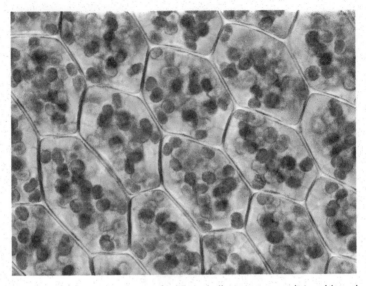

Figure 2.7 Microscopic view of chlorophyll, concentrated in chloroplast structures (source: Wikipedia).

When we introduced the mitochondrion, we promised to get back to the detailed role played by ATP, life's chemical "energy currency," as shown in Fig. 2.8.

ATP was discovered in 1929 by K. Lohmann. In 1935, V. Engelhart demonstrated that muscle contractions require ATP. C. Fiske and Y. Subbarow then made further contributions to our understanding of the role of Phosphorus (P) and P-containing molecules (such as ATP) in muscular activity. In 1948, A. Todd[14] was able to synthesize ATP. The next important step to be taken involved the role played by various enzymes.

Already in 1937, H. Kalckar established that ATP synthase is linked with cell respiration. In 1961, E. Racker and P. Mitchell[15] studied the mitochondrial membrane. Mitchell knew that the energy released in photosynthesis and cell respiration initiates a stream of positively charged hydrogen ions. Then he found that these hydrogen ions, in turn, drive the production of ATP with the aid of a membrane-bound enzyme called *ATP synthase*.

In 1964, Paul Boyer[16] proposed that ATP is synthesized through structural changes in the ATP synthase enzyme. He demonstrated that ATP synthase can be compared to a molecular machine, whose rotating bent axle is driven in a stepwise process by "biological electricity"—that is, the flow of hydrogen ions. Because of the asymmetry of the rotating axle, three subunits of the enzyme assume different forms and functions: a first form (i) that binds adenosine diphosphate (or ADP) and phosphate building blocks, a second form (ii) where these two molecules are chemically combined into a new ATP molecule, and a third form (iii) where the ATP that has been formed is released. In the next twist of the axle, the three subunits switch form and thus also function with each other, and another ATP molecule can be formed, and so on.

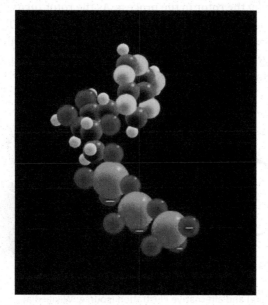

Figure 2.8 Molecular model of ATP (adenosine triphosphate), with hydrogen (white), oxygen (red), three phosphorous atoms (orange), nitrogen turquoise, and carbon black, and with ions marked by a negative (−) sign (source: Wikipedia).

The way ATP synthase is doing its job is one of the great miracles of evolution, a wonder of the molecular world. The ATP synthase enzyme is a molecular motor, an ion pump, and another molecular motor, all wrapped together in one nanoscale machine. There are two rotary motors, each powered by a different fuel. The motor at one end, frequently termed F0 (purple color in Fig. 2.9), is an electric motor powered by the flow of hydrogen ions across the membrane.

As the protons flow through the motor, they turn a circular rotor which is connected to the second motor, usually termed F1 (depicted in green color). The F1 motor is a "chemical motor," powered by ATP. Since the two motors are connected, it follows that "when F0 turns, F1 turns too."

Using an analogy to the human invention of electrical motors and generators, the F0 motor uses the power from a proton gradient to force the F1 motor to generate ATP. In our cells, food is broken down and used to pump hydrogen ions across the mitochondrial membrane. The F0 portion of ATP synthase allows these ions to flow back, turning the rotor in the process. As the rotor turns, it turns the axle and the F1 motor becomes a generator, creating ATP.

This *rotational catalysis* model was put forward by Boyer in the late 1970s (see Fig. 2.9), but only in 1994 did his ingenious "engineering model"[17] gain general acceptance among researchers.

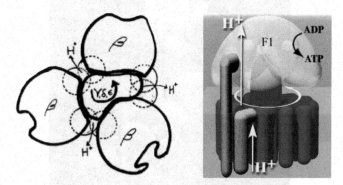

Figure 2.9 Rotational catalysis (adapted from Boyer's Nobel Lecture, 1997, Nobelprize.org) and alternative illustration (source: Wikipedia).

In August of 1984, Walker[18] and his colleagues published three-dimensional images of ATP synthase that had been obtained by x-ray analysis of enzyme crystals. Already 1981, Walker had determined the DNA sequence of the genes encoding the proteins in ATP synthase, but his x-ray images were needed to finally confirm Boyer's theory of ATP formation.

ATP synthase is present in abundance in intracellular membranes of mitochondria, plant chloroplasts, bacteria, and other organisms. The ATP is then transported out of the mitochondria and used for various body functions, including muscle, brain, nerve, kidney, liver, and other tissues. The ADP and phosphate formed when ATP is "used," return to the mitochondria where ATP is remade using the energy from oxidations. This process, also referred to as *oxidative phosphorylation*, is the most prevalent chemical reaction that occurs in all living organisms. It is estimated that organisms typically consume their body weights of ATP over the course of a day, i.e., each ATP molecule is recycled over 500 times in each body cell. ATP cannot be stored; hence its consumption closely follows its synthesis. At any time, ATP used by cell processes in the human body will add up to about 5 g.

The key role played by ATP synthase proves our initial point that proteins do the "heavy lifting," because enzymes are proteins. What we call oxidation of the food we eat is another set of chemical reactions that are catalyzed by various enzymes. Inside the cells, living organisms essentially burn the food, using oxygen and producing carbon dioxide and water. Oxygen is carried by the hemoglobin of red blood cells, then it reaches the mitochondria where it oxidizes other enzymes in a respiratory chain.

Let us turn to another important use of ATP, facilitated by yet another enzyme. In 1957, J. C. Skou[19] discovered "sodium, potassium ATPase (or Na+, K+-ATPase)," the enzyme which maintains the right ion balance in living cells. This mechanism is often referred to as a biological "pump" that transports potassium ions into a cell. On the other hand, sodium ions are transported in the opposite direction, out of the cell. This "biological pump" is a process requiring a lot of energy. Actually, up to one third of the ATP formed in the human body is used to drive the Na+, K+ pump! Today, we are also aware of a number of other ion pumps, which have been discovered as a consequence of Skou's pioneering work. All these ion pumps are prerequisites for various important life functions that include transmission of nerve impulses, muscle contraction, and digestion. Many pharmaceutical drugs, such as heart and ulcer medicines, are targeting the action of cellular ion pumps.

When defining "metabolism" above, we learned that living organisms need a way to *dispose* of waste.

On the cellular level, "waste disposal" means to get rid of unwanted proteins that may have played a useful role but are no longer needed. In eukaryotic cells, an organelle named *lysosome* was identified by C. de Duve[20] to perform important functions in "decomposing different types of materials, such as bacteria and parts of cells that have worn out." In 2004, the chemists A. Ciechanover, A. Hershko, and I. Rose were awarded another Nobel Prize for elucidating the detailed mechanisms involved in protein degradation in cells. They found that proteins destined for destruction are labeled with a 76-amino-acid-long (peptide) molecule that was first isolated in 1975, but later found in numerous different tissues and organisms— and therefore called "ubiquitin" (from Latin *ubique*, "everywhere"). Ubiquitin is shown in Fig. 2.10a.

The tagging of proteins is performed by special enzyme families named E1, E2, and E3, respectively. There are several hundred E3 enzymes, and they determine which proteins in the cell are to be marked and labeled with ubiquitin.

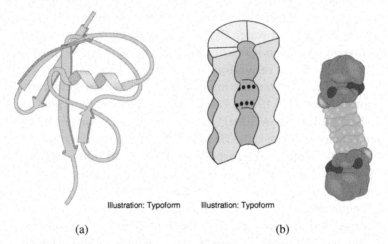

Illustration: Typoform   Illustration: Typoform

(a)   (b)

Figure 2.10   (a) Ubiquitin, a molecular label that represents the "kiss of death." (b) The proteasome, the cell's waste disposer (source: Wikipedia).

Ubiquitin is then detected by another cellular "machine" named proteasome and depicted (in green color) in Fig. 2.10b. Inside the proteasome, the ubiquitin labels are disconnected, the protein is admitted and then chopped into small pieces, about seven to eight amino acids long. Those peptide pieces can then be further degraded into shorter amino acid sequences and used in synthesizing new proteins. We know now that "ubiquitin activation" and "controlled protein breakdown" are necessary functions in a healthy cell, for the cell to function and reproduce properly. Malfunctioning is suspected to lead to human diseases such as neurodegenerative disorders (including Alzheimer's disease), cancer, and cardiovascular problems.

To understand how life on Earth evolved from simple prokaryotes to complex multicellular organisms, we finally need to discuss *evolution*.

Charles Darwin defined *natural selection* as the differential survival and reproduction of individuals due to differences in phenotype. That is a key mechanism of *evolution*, the change in heritable traits of a population over time. Natural selection acts on the phenotype, but the genetic (heritable) basis of any phenotype that gives a reproductive advantage, provides the underpinning. Over time, this process can result in populations that adapt to environmental conditions ("microevolution") but may eventually result in the emergence of new species ("macroevolution"). On a molecular level, changes in the genome of an organism are caused by random DNA mutations. Herman Muller[21] studied the hereditary characteristics of fruit flies. In 1927, he discovered that the number of genetic mutations increased when fruit flies were exposed to x-rays. He found that the number of mutations were related to the *dose* of x-rays and other ionizing radiation the fruit flies were exposed to. By intervening in what was previously called "the blind play of Nature," he was able to induce mutations that were reflected as changes in phenotypic properties. Muller thus opened up new possibilities for experimental investigations and pointed the way towards a molecular understanding of Darwin's fundamental insight. More than 90 years later, as will be covered at the end of this book, we have found a new understanding of genotypic-phenotypic connections and learned new ways to edit and modify genetic traits, leading both to scientific breakthroughs and bioethical challenges.[22]

Charles Darwin was right in assuming that throughout the lives of individuals, their genomes interact with their environment to cause variations in traits, which can be inherited by their offspring.

Evolution has been driving the changing life forms on Earth until today, as summarized in Figs. 2.5a and 2.5b. That includes all living organisms, from archaea and bacteria to homo sapiens.

Figure 2.11   Charles Darwin, aged 51 (source: Wikipedia).

An interesting initiative currently discussed by the scientific community is the Earth BioGenome Project (EBP). Conceived by Harris Lewin, an evolutionary genomicist at the University of California at Davis, the EBP would take a step toward its audacious goal to sequence all life forms on the planet Earth, by first focusing on eukaryotes. According to a preliminary plan, Step One would be to create a "reference genome" for each of about 9000 eukaryotic families. In Step Two, the EBP team would select a species from each of the 150,000 to 200,000 genera for sequencing. Finally, EBP participants would get rough genomes of the 1.5 million remaining known eukaryotic species. Commitments to contribute to EBP with sequencing equipment and other resources have so far been obtained from China's BGI, Shenzhen, and from the Wellcome Trust via its Sanger Institute in the United Kingdom.[23]

Chapter 3

Adaptation of Life to Extreme Conditions

What makes life on the planet Earth so diverse and interesting is the existence of extreme conditions, such as hot and cold temperatures, dry deserts and deep waters, hot volcanos and icy glaciers, low elevations with oxygen-rich air and high elevations making it very difficult to breathe normally.

In addition, temperatures may vary wildly between hot summers, temperate springs and autumns, and frosty winters.

When we introduced archaea, we already mentioned that those microorganisms are able to survive in harsh environments under extreme conditions.

What is more astonishing is that complex life forms such as mammals, birds, reptiles, and fish are also surprisingly adaptable.

Zoologists have tried for centuries to classify animals. The concept of *clade* has survived as a way to divide mammals, birds, and reptiles. A clade can be defined as a large group of organisms that consists of a common ancestor and represents a single "branch" on the "tree of life."

Mammals are a clade (taxon *Mammalia*) and birds, too, are a clade, formally referred to as the taxon *Aves*. Mammalia and Aves are, in fact, *subclades* covered by the higher level clade of *Amniota*. They are defined by the fact that they lay their eggs on land or retain the fertilized egg within the mother. Amniota are distinguished from the

Our Animal Connection: What Sapiens Can Learn from Other Species (Second Edition)
Michael Hehenberger and Zhi Xia
Copyright © 2021 Jenny Stanford Publishing Pte. Ltd.
ISBN 978-981-4877-50-3 (Hardcover), 978-1-003-13072-7 (eBook)
www.jennystanford.com

anamniotes (fishes and amphibians), which typically lay their eggs in water.

Surprisingly, Reptilia do not form a separate clade. They belong to the clade Amniota and are defined by being amniotes that lack fur or feathers. One could say that the traditional Reptilia are "non-avian, non-mammalian amniotes."[24]

Mammals are vertebrates (possessing a hollow back side or *dorsal* nerve cord, and backbones) and are "warm-blooded" or endotherm, meaning that they maintain their body at a metabolically favorable temperature, largely by the use of heat set free by the animal's routine metabolism. The dorsal nerve cord is modified into the central nervous system, which comprises the brain and spinal cord. Mammals are also distinguished by the possession of a neocortex, which is defined as the part of the brain involved in sensory perception, cognition, generation of motor commands, spatial reasoning, and language. Mammals have hair or fur, three middle ear bones, and mammary glands that are used by female mammals to nurse their young with milk.

Although the basic body type is a terrestrial quadruped, some mammals are adapted for life at sea, in the air, in trees, and even underground. The largest group of mammals are the "placentals," i.e., the females have a placenta, which enables the feeding of the fetus during gestation. Mammals range in size from the 30 to 40 mm *bumblebee bat* to the 30-meter *blue whale*, the largest animal on the planet.

Most mammals are intelligent. They possess large brains, have self-awareness, and are able to use tools. Mammals can communicate and vocalize in several different ways. They can produce all kinds of sounds, including the production of ultrasound, singing, and echolocation. Some communicate via scent-marking and alarm signals. Mammals can be both solitary and territorial, but also able to organize themselves into various types of societies and hierarchies.

After 3 billion years of evolution, life can be found in every corner of the earth. From the depths of the ocean, the ocean floor to the upper stratosphere, signs of life activity can be detected. From hot and cold temperatures, high to low oxygen environments, to pure acid and atomic bomb radiation, there seems to be no extreme environment that the well-adapted organism cannot withstand. Understanding the evolutionary genetic mechanism of organisms' adaptation to the environment in extreme environments provides a new way for systematically and globally revealing the genetic mechanisms of natural selection.

How are mammals dealing with extreme conditions such as long, cold winters? They have three basic strategies for survival in winter: they can hibernate, they can migrate, or they can stay put and tough it out. Cold temperatures may be difficult to deal with, but lack of food is generally considered to be the primary factor that causes many species (and not only mammals) to migrate or hibernate.

Although cold climate leads to reduced levels of activity, there's still a lot going on in winterland—both above and below the snow. Many animals are able to remain active since they have adapted to winter's challenges.

Animals use fat as insulation to keep their bodies warm in winter. These layers of fat act as an internal coat and help provide energy when food is scarce during the winter months. Specialized fat, called brown fat, is produced during the food-rich seasons and expended during winter. Brown fat is used by most hibernators for arousal and by many migrators as a fuel source.

What is Hibernation?

Hibernation is a state of inactivity and *metabolic depression*, a way of "throttling down all the body's engines." The state of hibernation is characterized by low body temperature, slow breathing and heart rate, and low metabolic rate. Active metabolic suppression is a way to deal with low temperatures, down to just above the freezing point. Hibernating mammals are able to reduce heartbeats to only 3–5 per minute, without suffering from lethal blood clots. When waking up from weeks or months of hibernation, normal clotting activity is quickly restored.

Before entering hibernation, animals need to store enough energy to last through the entire winter. Larger species eat a vast amount of food and store the energy in fat deposits. Many small species use another strategy, food caching. Some species of mammals even manage to give birth to young ones while hibernating. In addition, they are able to restore full organ function after e.g. having almost completely shut down their kidneys during hibernation.

Having discussed adaptation to low temperatures in winter, what about the opposite, withstanding heat? The species showing perhaps the most spectacular adaptation to high temperature is the Pompeii worm, here mentioned as a curiosity although not being a member of the mammalian clade. *Alvinella pompejana* is a species of deep-sea polychaete worm.[25] It is an extremophile found only at hydrothermal vents in the Pacific Ocean. The Pompeii worm is adapted to the high temperatures and pressures of the hydrothermal vents by using a thick layer of bacteria to protect it from heat, and by hiding inside a papery tube to protect it from predators. Its tail ends are often resting in temperatures as high as 80°C! On the other end, the worm's feather-like head sticks out of the tubes into water that is a much cooler, around 22°C.

Birds are a group (clade) of endothermic vertebrates, characterized by feathers and toothless beaked jaws. Reverse genetic engineering and the fossil records both demonstrate that birds are modern feathered dinosaurs, having evolved from earlier feathered dinosaurs.

They are laying hard-shelled eggs, have a high metabolic rate, and a strong yet lightweight skeleton. As discussed in more detail below, they have a very efficient way of oxygenating their blood via respiration. They beat mammals in efficiency. Birds can be found everywhere and range in size from the 5 cm *bee hummingbird* to the 2.75 m *ostrich*. They rank as the world's most numerically successful class of tetrapods, with approximately ten thousand living species. Birds have wings which are more or less developed depending on the species. Wings, which evolved from forelimbs, gave birds the ability to fly, although further evolution and adaptation has led to the loss of flight in some species, such as penguins. The digestive and respiratory systems of birds are also uniquely adapted for flight. Some bird species of aquatic environments, particularly seabirds and some waterbirds, have further evolved for swimming.

Life on Earth is completely dependent on oxygen and most living organisms require a continuous and uninterrupted supply of oxygen to survive.

Respiration is therefore a very important activity for all animals and also subject to adaptation.

In terrestrial mammals, the respiratory surface is internalized as linings of the lungs. In mammals and reptiles, the gas exchange in the lungs occurs in millions of small air sacs called *alveoli*. In birds, the air sacs are called *atria*.

The function of these microscopic air sacs is to bring the oxygen of the air into close contact with the blood. They communicate with the external environment via a system of airways, or hollow tubes, of which the largest is the trachea. It branches into the two main bronchi which continue to further branch into narrower secondary and tertiary bronchi, and finally the *bronchioles,* as shown in Fig. 3.1.

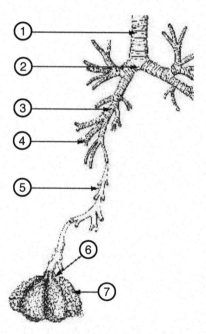

Figure 3.1 Lower respiratory system in mammals: (1) trachea, (2) mainstem bronchus, (3) lobar bronchus, (4) segmental bronchus, (5) bronchiole, (6) alveolar duct, (7) alveolus (source: Wikipedia).

In birds the bronchioles are termed *parabronchi*. Figure 3.2 shows the avian respiratory system in some detail.

The respiratory system of birds differs significantly from that found in mammals. They have rigid lungs which do not expand and contract during the breathing cycle. An extensive system of air sacs distributed throughout their bodies act as the bellows drawing environmental air into the sacs, and expelling the spent air (mostly carbon dioxide) after it has passed through the lungs. Bird lungs are smaller than those in mammals of comparable size, but the air sacs account for 15% of the total body volume, significantly higher than the only 7% devoted to the above-mentioned alveoli of mammals. Because of this advantage, birds are extremely efficient in their use of oxygen, even at high altitude. No wonder that the highest measured altitude reached by an animal was the accidental Rüppell's vulture collision with a jet engine in Africa, at 11,300 meters.[26]

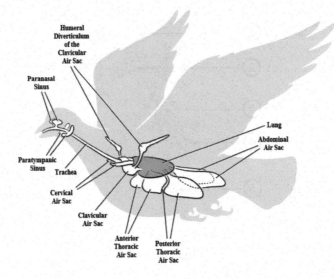

Figure 3.2 Arrangement of air sacs and lungs in birds (source: Wikipedia).

Migration is another way to adapt to global differences in temperature and climate. Many bird species are migrating to optimize their food sources and to ensure the survival of their offspring. Landbirds, shorebirds, and waterbirds undertake annual long-distance migrations, usually triggered by the length of daylight as well as weather conditions. They usually spend the breeding season in the temperate or polar regions and the rest of the year in the tropical regions, in some cases even in the opposite hemisphere. Before migration, birds substantially increase body fats and reserves and reduce the size of some of their organs.

Migration is highly demanding, in particular as birds need to cross deserts and oceans without refueling. Landbirds have a flight range of around 2500 km and shorebirds can fly up to 4000 km. A female bar-tailed godwit, a large, streamlined shorebird, has been tracked via satellite transmitter and shown to perform a nonstop flight of as far as 10,200 km![a] Seabirds are also able to undertake long migrations. The longest annual migration among seabirds has been recorded by sooty shearwaters.

As shown in Fig. 3.3, the Arctic tern seems to hold the overall long-distance migration record for birds, traveling between Arctic breeding grounds and the Antarctic each year. Some species of tubenoses (Procellariiformes) such as albatrosses circle the Earth, flying over the southern oceans, while others such as Manx shearwaters migrate 14,000 km between their northern breeding grounds and the southern ocean. Shorter migrations are common, including altitudinal migrations (e.g., by geese) on mountains such as the Andes and Himalayas.

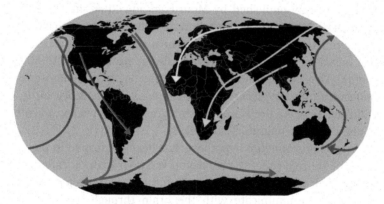

Oenanthe oenanthe		Northern Wheatear
Sterna paradisaea		Arctic Tern
Falco amurensis		Amur Falcon
Puffinus tenuirostris		Short-tailed Shearwater
Philomachus pugnax		Ruff
Buteo swainsoni		Swainson's Hawk

Figure 3.3   Bird migration routes (source: Wikipedia).

---

[a]https://www.sciencedaily.com/releases/2007/09/070915131205.htm

A big question related to bird migration is *bird navigation*. How do birds (and some other species like turtles and butterflies) find their way when migrating thousands of kilometers, across oceans without help of landmarks? What are the navigation tools those migrating birds are using?

This question has occupied scientists and philosophers since ancient times. Are migrating species somehow able to use the Earth's magnetic field or are they using the positions of the sun (during the day) and the stars (at night) to navigate? Although there is still research going on, what we know today is that *magnetoreception* plays a key role and that a deep explanation of the navigation mechanisms used requires an understanding of quantum theory. What is further obvious is that humans are at a disadvantage: While birds have built-in navigation, sapiens had to invent the compass to find the Northern direction, had to use sun, moon, and stars for guidance, and to eventually send rockets into space to develop satellite-based navigation devices.

A good review of magnetoreception and its use in bird navigation was published by Mouritsen and Ritz[27] in 2005. They summarized their findings as follows: "Many birds seem to have two magnetodetection senses, one based on magnetite near the beak and one based on light-dependent radical-pair processes in the bird's eye(s)." They further pointed to the existence of "magnetosensory molecules," the *cryptochromes*, in the eyes of migratory birds. In addition, there is a brain area that integrates specialized visual input at night in night-migratory songbirds.

The *beak magnetite* hypothesis has been confirmed in pigeons: a magnetite-rich (microscopic iron crystal) structure located in the upper beak of pigeons has been found to be connected to nerve fibers that communicate with the brain through the trigeminal nerve.[b] Experiments have shown that pigeons are able to detect a strong magnetic anomaly. However, it is doubtful that biologically relevant small changes in magnetic field strengths can be detected this way.

The second magnetodetection hypothesis, often named "chemical magnetodetection" or "light-dependent magnetoreception" has been shown to provide the magnetic compass of night-migratory

---

[b]A cranial nerve responsible for sensation in the face and motor functions such as biting and chewing.

songbirds and of homing pigeons. It has further been demonstrated that visual photoreception is the underlying mechanism. On a molecular level, blue-to-ultraviolet (UV) light stimulates a protein called cryptochrome 1a (CRY1a) that is present in the birds' retinal photoreceptor cone cells. When struck by light, CRY1a changes into one of two states that only differ in the position of an unpaired electron. Since molecules with unpaired electrons are called "radicals," both states are referred to as CRY1a "free radicals." The ratio of these two states depends on the orientation of cryptochrome to magnetic fields and is highly sensitive to even small changes of magnetic field strength. Therefore, the "radical-pair" theory of bird navigation seems to now have won over most scientists working on this problem.

The conclusive result[28] of decades of bird navigation studies appears to be that birds rely primarily on magnetodetection—they navigate chemically with cryptochrome molecules in the retina of their eyes, and visually by tracking the position of the sun and stars, with perhaps some additional support of beak magnetites.

Reptiles are tetrapod animals in the class Reptilia, comprising today's turtles, crocodilians, snakes, lizards, etc., along with many extinct relatives. Because some reptiles are more closely related to birds than they are to other reptiles—surprisingly, crocodiles are closer to birds than to lizards—the traditional groups of "reptiles" listed above do not together constitute a clade, i.e., "consisting of all descendants of a common ancestor."

The earliest known proto-reptiles, living under water, originated around 312 million years ago. They later adapted to life on dry land. Modern non-avian reptiles inhabit all the continents except Antarctica. As to special adaptations, reptiles show a lot of diversity. As will be discussed later, they are masters of disguise and regeneration.

Snakes have another way to slow down their metabolism, called *brumation*. Brumation is an example of dormancy in reptiles that differs from hibernation in the metabolic processes involved. When mammals hibernate, they are actually asleep. When reptiles brumate they are far less active; they do not eat, but are still required to drink water. Brumation is triggered by lack of heat and the decrease in the hours of daylight in winter, similar to hibernation. Snakes often gather together for brumation in large numbers (sometimes over 1000 snakes), huddling together inside underground dens.

Another way of adapting to extreme conditions is practiced by *marine mammals*, including seals, whales, manatees, sea otters, and polar bears. They rely on the ocean and other marine ecosystems for their existence. They are unified by their reliance on the marine environment for feeding and have developed a number of ways to overcome the challenges associated with aquatic living. Marine mammals have achieved efficient locomotion via torpedo shaped bodies to reduce drag; their limbs are modified for propulsion and steering; some have even developed tail flukes and dorsal fins for propulsion and balance.[29] Marine mammals are further adapted to very cold temperatures: polar bears, otters, and seals have developed furs that are long, oily, and waterproof to trap air and provide insulation. Others, such as whales, dolphins, manatees, and walruses, have traded long fur in favor of a thick, dense epidermis and a thickened fat layer (blubber) in response to aquatic requirements

As to their respiratory systems, marine mammals are able to dive for long periods of time while relying on large and complex blood vessel systems to store oxygen. Some have developed oxygen reservoirs inside their muscles, blood, and the spleen, all having the capacity to hold a high concentration of oxygen. Marine mammals are also capable of bradycardia (reduced heart rate), and vasoconstriction—a way of restricting the use of oxygen to vital organs such as the brain and heart—to extend diving times and cope with oxygen deprivation. Even if oxygen is depleted, marine mammals can still access substantial reservoirs of glycogen that support anaerobic glycolysis of the cells involved during hypoxic conditions.

Sound travels differently through water, and therefore marine mammals have developed adaptations to ensure effective communication. Their most notable adaptation is the development of echolocation in whales and dolphins.

After our discussion of marine mammals, let's turn to **fish**. They represent 53% of all vertebrates, more than birds, reptiles, and mammals taken together. About 32,000 species of fish have been found on Earth, with the oceans accounting for two-thirds of the population. The rest live in fresh water. Fish and humans have gone through more than 5000 years of shared history and have formed a close relationship, becoming an extremely important food and ornamental pet in human daily life. But which animals belong to "fish"? This is a difficult question to answer, because fishes have a huge variety of body forms. More than 2400 years ago, the Greek philosopher Plato defined fish as living organisms created by the Lord as "the descendants of all kinds of sins and unclean hearts." Plato believed that fish were put into water as a punishment, to force them "to breathe the wonderful and pure air through the deep sludge."

Fortunately, Plato's creationist view has been completely negated by science. The modern definition of fish goes as follows: fishes are animals that are cold-blooded (except for tunas and marlins and mako sharks that are warmer than the water), breathe using gills (although there are exceptions: lungfish and some others have lungs), have a backbone or a notochord (but not always, such as in sharks, which rely on cartilage, a tough material that composes a shark's skeleton), have a scaly skin (except for eels, which are scaleless), and have various fins instead of limbs (except for a few that do actually have limbs, like lungfish and coelacanth).[30]

It is interesting that the hearing of fish is really good. Scientists have found that although many fish do not have long ears outside, there are specially designed sound receivers that can transmit sound waves to the liquid-filled tubular structure in the inner ear. These pipes have special fine hairs, called cilia, that can transmit sound pulses through a series of complex mechanisms and chemical reactions to the brains of the fish where they are processed. The *otolith* is part of the auditory system and is connected to sensory cells and plays a major role in the auditory/balancing mechanism of teleosts.[c] Scientists rely on otoliths to discriminate the type of fish. Via otoliths it is further possible to determine the age of a fish, because the otolith grows a concentric circle each year as the fish develops. Under the microscope, scientists can see and count these concentric circles. Fish vision is also highly developed. Nearly all daylight fish have color vision. Many fish species also have chemoreceptors that are responsible for extraordinary senses of taste and smell. Most fish have sensitive receptors that form the lateral line system (consisting of an array of sensors called neuromasts along the length of the fish's body), which detects gentle currents and vibrations, and senses

---

[c]The teleosts belong to the ray-finned fishes and make up 96% of all current fish species.

the motion of nearby fish and prey. For example, sharks can sense frequencies in the range of 25–50 Hz through their lateral line. Fish orient themselves using landmarks and may use mental maps based on multiple landmarks or symbols. Fish behavior in mazes reveals that they possess spatial memory.

A typical fish has the following basic body plan,[31] as shown in Fig. 3.4:

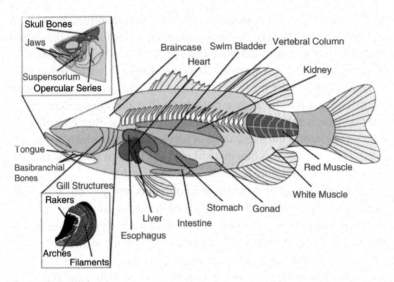

Figure 3.4 Body plan of fish including gill structures (source: Wikipedia).

Throughout history, humans have utilized fish as a food source. Historically and today, most fish protein has come by means of catching wild fish. However, aquaculture, or fish farming, is becoming increasingly important in many nations. Overall, about one-sixth of the world's protein is estimated to be provided by fish. The most popular species include herring, cod, anchovy, tuna, flounder, and salmon. Fish have also long been recognized as objects of beauty. They appear in cave art, are raised as ornamental fish in ponds, and are displayed in aquariums.

It is interesting how gills extract dissolved oxygen from water and excrete carbon dioxide. In fish (and mollusks such as octopus and squid), the efficiency of the gills is enhanced by a countercurrent mechanism: the water passes over the gills in the opposite direction to the flow of blood through them. This mechanism is very efficient and permits recovery of about 90% of the dissolved oxygen in water.[32] The gills are composed of comb-like filaments, the gill lamellae, which help increase their surface area for oxygen exchange.

A good example of special fish adaptation is provided by the *Greenland shark*. Like most shark species, it is almost immune to all known diseases, and it may be able to live for more than 400 years, due to growth rates and biochemical activities that are slowed down by cold water. Studies have shown that Greenland sharks are the longest living vertebrates, reaching sexual maturity as late as at the age of 150 years.[33]

Similar to birds, some fish species can migrate across large distances. Salmons are known to regularly migrate to and from their breeding grounds. They spend their early life in rivers, and then swim out to sea where they live their adult lives and gain most of their body mass. When they have matured, they return to the rivers to spawn, usually exactly to the natal river where they were born, even to the very spawning ground of their birth. When they are in the ocean, they probably use *magnetoception* (detection of the Earth's magnetic field) to locate the general position of their natal river. Once close to the river, they probably use their sense of smell to find the river entrance and their natal spawning ground.

Note that magnetoreception is present in bacteria, arthropods, mollusks and a number of vertebrates, but not humans. Humans do not have a magnetic sense, but there is a protein (a cryptochrome) in the eye which could potentially serve this function.[34]

Finally, let's discuss the special challenges provided by extreme conditions deep down in our oceans. How can fish, birds, and some mammals adapt? Unlike terrestrial animals, they rely on oxygen stored in their muscles while holding their breath. For instance, *emperor penguins* can store 47% of total body oxygen in their muscles when they dive to a depth of up to 510 m. Scientists have also found that the content of myoglobin is extremely high among deep-sea master muscles belonging to dolphins, seals and sperm whales. Myoglobin is a protein that absorbs and stores oxygen in the blood. Therefore, even if these animals do not breathe, they can survive submerged for long hours.

Another trick practiced by deep diving animals is their ability to slow down their heart beats. By attaching sensors to walruses to measure their heart rate, it was found that they drop the average heart rate from 107 beats per minute down to only 39 beats per minute. In another test of elephant seals, the researchers even found a reduction to only 3 beats per minute.[35] A slower heart rate means a decline in metabolism. Therefore, these animals can still perform deep submerged predation for a long time in the presence of low oxygen levels and slow blood circulation.

When the deep-sea animals dive, the nitrogen concentration in the body remains basically unchanged. While humans and other poorly adapted mammals are suffering from decompression sickness due to the release of dissolved gases as bubbles, leading to serious health issues, deep-sea animals are able to avoid such problems.

The terminology used to describe the deep-sea environment is shown in Fig. 3.5. Note that there are even sweet water environments extending beyond a depth of 1600 m in Lake Baikal, Siberia. The greatest ocean depth is the Mariana Trench near Guam, at 10,911 m.

Since ancient times and until the 20th century, scientists believed that life at great depth was more or less nonexistent. The eternal dark, the almost inconceivable pressure and the extreme cold that exist below 1000 m were thought to be so forbidding as to permit no life. However, the reverse is in fact true: below 200 m lies the largest habitat on earth.[36]

Lanternfish, named after their use of bioluminescence (involving some light-emitting molecule and the luciferase enzyme), account for as much as 65% of all deep-sea fish biomass.[37] They are among the most widely distributed, populous, and diverse of all vertebrates, playing an important ecological role mostly as prey for larger organisms. Their estimated global biomass is much greater than what is caught by all the fisheries in the entire world. Lanternfish are well known for their vertical migrations: during daylight hours, they remain at 300 and 1500 m depth, mostly within the gloomy bathypelagic zone. Towards sundown, they then rise into the epipelagic zone, between 10 and 100 m deep. They probably do this both to avoid predation and because they are following the vertical migrations of their food—zooplankton (i.e., mostly microscopic life forms).

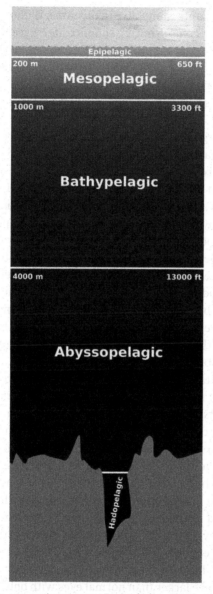

Figure 3.5 Deep-sea environments. Zones below 1000 m are aphotic (no light!) (source: Wikipedia).

With the exception of the upper parts of the mesopelagic, there is no natural light that penetrates the deep sea. Since photosynthesis is not possible, plants cannot live in this zone. Without plants, where does the energy come from to support life? Except for the areas close to hydrothermal vents, it actually comes from organic material drifting down from the photic zone. The sinking organic material is collectively referred to as "marine snow."

Because pressure in the ocean increases by about 1 atmosphere for every 10 m of depth, the amount of pressure experienced by many marine organisms is extreme. Traps that incorporate a special pressure-maintaining chamber, are required to examine animals that have been retrieved from the deep sea.

At any given depth, the temperature is practically constant over long periods of time. There are no seasonal temperature changes, nor are there any annual changes. No other habitat on earth has such a constant temperature. Only hydrothermal vents are the exception, there the temperature of the water as it emerges from the "black smoker" chimneys may be as high as 400°C. It is kept from boiling by the high hydrostatic pressure. Within a few meters from the vent it may be back down to 2–4°C. Organisms living close to such vents are also very interesting, they are examples of life on earth not dependent on the sun. They obtain nutrients and energy directly from thermal sources and from chemical reactions associated with mineral deposits. These organisms depend on hydrogen sulfide, a compound that is highly toxic to almost all terrestrial life. The fact that life can exist under these extreme conditions may open up the possibility that there may be life elsewhere in the universe.

Let's return to midwater and deep-sea fish and their special adaptations. They tend to be small, usually measuring less than 25 cm. They have slow metabolisms and unspecialized diets, preferring to sit and wait for food rather than searching for it. They have elongated bodies with weak, watery muscles and skeletal structures. They have extendable, hinged jaws in their big mouths. Finding a partner with which to breed is difficult, so many organisms are hermaphroditic. Because of the low or nonexistent light, deep-sea fish often have larger than normal eyes with only rod cells. Their upward field of vision allows them to seek out the silhouette of possible prey. To cope with predation, their adaptations are mainly concerned with reduction of silhouettes, as a form of camouflage. They reduce the area of their shadow by lateral compression of the body, and they counter illumination via bioluminescence.

Fish living in the deep sea naturally evolved a new internal structure through the molecular and cellular levels to counter the high-pressure environment. To adapt, deep-sea fish cell membranes have to become more fluid. To withstand the enormous water pressure in the deep-sea environment, their bones are thin and easy to bend; the muscle tissue becomes particularly flexible and the fibrous tissue becomes surprisingly fine. Deep-water fish can no longer live in shallow water. Once they reach shallow water, the internal pressure would cause all cell content to seep out.

Animals are defined as multicellular eukaryotic organisms that form the biological kingdom Animalia. With few exceptions, animals consume organic material, breathe oxygen, are able to move, reproduce sexually, and share a basic embryonic development that starts from a hollow sphere of cells, the blastula. Over 1.5 million living animal species have been described—of which around 1 million are insects—but it has been estimated that there are over 7 million animal species in total. Animals have complex interactions with each other and their environments, forming intricate food webs. The study of animals is called zoology.

Many modern animal phyla (a level of biological classification/taxonomic rank below *kingdom*, but above *class*) became clearly established in the fossil record as marine species during the Cambrian explosion which began around 542 million years ago. Over 6000 groups of genes common to all living animals have been identified and may have arisen from a single common ancestor that lived 650 million years ago.

Aristotle divided animals into those with blood and those without. Carl Linnaeus created the first hierarchical biological classification for animals with his Systema Naturae,[38] which Jean-Baptiste Lamarck[39] expanded into 14 phyla by 1809. In 1874, Ernst Haeckel[40] divided the animal kingdom into the multicellular Metazoa (now synonymous with Animalia) and the Protozoa, single-celled organisms no longer considered animals, but still covered in this book whenever there is a connection to human health. Today's biological classification of animals relies on advanced techniques, such as molecular phylogenetics. Increasingly, scientists are critically examining the conventional evolutionary relationships between animal taxa.

Arthropods (from Greek arthron = joint and pous = foot) are invertebrate animals having an exoskeleton (external skeleton), a segmented body, and paired jointed appendages. Arthropods include, among others, insects and crustaceans. They are characterized by their jointed limbs and cuticle made of chitin,[d] often mineralized with calcium carbonate. The arthropod body plan consists of *segments*.

Their versatility has enabled them to become the most species-rich members of all ecological guilds in most environments. They have over a million described species, making up more than 80 per cent of all described living animal species. Arthropods have an internal cavity accommodating their internal organs. Instead of blood they circulate "haemolymph"[e] in their open circulatory systems. The internal organs of arthropods are built of repeated segments and their nervous system is "ladder-like," with paired ventral nerve cords running through all segments and forming paired ganglia in each segment.

Their heads are formed by fusion of varying numbers of segments, and their brains are formed by fusion of the ganglia of these segments. Arthropods have respiratory and excretory systems that vary with their environment.

Arthropod species have various forms of vision and other types of perception, from very simple to highly advanced. Various combinations of compound eyes and pigment-pit ocelli can be found: in most species the ocelli, defined as simple eyes consisting of retinal cells, pigments, and nerve fibers, can only detect the direction from which light is coming, and the compound eyes are the main source of information, but the main eyes of spiders are ocelli that can form images and, in a few cases, can even swivel to track prey. Arthropods also have a wide range of chemical and mechanical sensors, sitting on bristles that project through their cuticles.

Their methods of reproduction and development are also interesting: all terrestrial species use internal fertilization, while aquatic species use either internal or external fertilization. Almost all arthropods lay eggs, but scorpions give birth to live young. Some arthropod hatchlings undergo a total metamorphosis to produce the adult form. The level of maternal care for hatchlings can be nonexistent or—again with scorpions—quite intense and prolonged.

---

[d]Chitin is the main ingredient in the exoskeletons of arthropods and crustaceans.
[e]A fluid equivalent to blood in most invertebrates, occupying the hemocoel.

Cell Biology and Genomics of Adaptation

At the biochemical level, virtually all aspects of cell function and metabolism can be the target of adaptive change in response to environmental stress. For readers familiar with details of cell biology, we list a number of mechanisms associated with adaptation, as described in more detail by Ken Storey et al[41]:

- Increasing or decreasing the expression of selected genes to produce corresponding changes in the levels of various proteins
- Elaborating novel genes and proteins that address stress-specific concerns
- Altering the kinetic and regulatory properties of enzymes and functional proteins
- Changing enzyme susceptibility to post-translational modification, in particular the effects of reversible phosphorylation by protein kinases and protein phosphatases
- Modifying sensing and signaling mechanisms by changes to cell surface receptors, signal transduction cascades, cross-talk between signaling pathways, and the targets of signals including proteins, transcription factors, and genes
- Changing protein–protein binding interactions that alter the composition of enzyme/protein complexes or localize enzyme/protein function via associations with specific binding proteins or subcellular structures
- Changing membrane composition to compensate for changes in environmental factors including temperature, pH, and ionic composition
- Producing protective molecules that can defend cell volume and/or stabilize protein or membrane structure/function

Finally, here are some examples of adaptation to extreme conditions where genomic studies have helped us to improve our understanding of some animals' adaptation mechanisms.

Belgica Antarctica: The Antarctic maggot is a small, wingless insect whose larval stage lasts for two years and is mostly frozen under the ice in the Antarctic. The insects mate and spawn within seven to ten days of adulthood and then die. Interestingly, the Antarctic cricket genome contains only 99 million bp (base pairs) of nucleotides; it lacks the non-coding DNA sequences and repetitive elements found in the genomes of most animals. The Antarctic genomic selection probably chose to abandon these "negative burdens" in order to adapt to extreme conditions and to survive in the Antarctic dry and cold conditions.[42] Although the Antarctic pupa genome is small, the number of genes is similar to that of other mosquitoes and contains approximately 13,500 functional genes. Most animals start heat shock proteins (especially high or low temperature conditions) when extreme pressure is encountered, and the pressure is turned off soon after the pressure is applied. However, these proteins are continuously activated during the larval stages of Antarctic mites. This fact is also considered to be related to their ability to withstand harsh environments. It should further be noted that, while most insects can survive 20% water loss, Antarctic grasshoppers can tolerate as much as 70% of water loss.

Yunnan Snub-Nosed Monkey: The Tibetan Plateau has an average elevation of more than 4000 m. Most mammals (including humans) experience problems there because, as the altitude rises, the oxygen partial pressure will decrease. As a consequence, mammals experience breathing difficulties and high-altitude sickness. So what is the secret of the golden monkey's adaptation to the high-altitude environment of the Tibetan Plateau? On the basis of multilevel studies, including genetic and sequence analysis of species and populations, transcriptomes and functional experiments, researchers[43] discovered genetic mechanisms associated with the adaptation of golden monkey species to high-altitude environments: the genes of gilt monkeys were significantly enriched in DNA repair and oxidative phosphorylation. Both the CDT1 gene, which is known to be associated with DNA repair, and the angiogenesis-associated RNASE4 gene have developed stable mutants that help golden monkeys adapt to high-altitude environment.

Metagenomics and Extreme Condition Adaptation: Another study[44] performed at the extreme environment of the Qinghai-Tibet Plateau has focused on complex traits such as tolerance to hypoxia, extreme cold, and other rough conditions. In addition to genetic mechanisms for adaptation in the genetic material of animals themselves, the investigators studied contributions of the gut microbiome (see a full chapter below in this book) that had co-evolved with animal hosts. Based on the hypothesis that adaptation to the long-term extreme environment of the plateau would require "efficient energy generation and utilization" and "low energy loss," scientists expected that short-chain fatty acids would be the main source of energy, and measured methane production, a by-product of rumen fermentation in digestive tracts. The study found that Yaks that were well adapted to the extreme environment of the plateau, produced significantly shorter-chain fatty acids than lowland yellow cattle, and that the yaks' methane emissions were significantly lower than those of yellow cattle. Through rumen microbial metagenomic sequencing analysis, it was found that the yak genes in the anabolic pathway of short-chain fatty acids were significantly enriched, and that yellow cattle genes were enriched in the methane formation pathway. By comparing the rumen mucosal transcriptomes of yak and cattle, the genes related to the transport and uptake of yak short-chain fatty acids were significantly upregulated, demonstrating the co-evolution of hosts and gut microbes, and further suggesting that plateau animals themselves can better absorb and utilize rumen fermentation. The study shows that co-evolution of the host and the microbiome in the process of adapting to the extreme environment of the plateau is one of the inevitable choices for mammals in the same region to adapt to evolution.

Chapter 4

Homo Sapiens ("Us"): Strengths and Weaknesses

The binomial name *Homo sapiens* was coined by the Swedish scientist Carl Linnaeus in 1758. The noun homo means "man, human being," and sapiens can be translated as "wise."

According to genetic and fossil evidence, the evolution of *Homo erectus* out of the last common ancestor of chimpanzees and humans happened roughly 10 to 2 million years ago. Later, Homo sapiens evolved from Homo erectus about 1.8 million to 200,000 years ago. It is widely believed that our species Homo sapiens developed into anatomically modern humans solely in Africa, between 200,000 and 100,000 years ago, with members of one branch leaving Africa by 60,000 years ago and over time replacing earlier human populations such as *Neanderthals* and *Homo erectus.* Only Homo sapiens survived, all other human species became extinct.

However, recent sequencing of the full Neanderthal genome[45] indicates overlap of up to 4% between sapiens and Neanderthal DNA, suggesting that cross-breeding led to Neanderthal genetic admixture. The mixing of sapiens and Neanderthal DNA cannot be found in Africa. This finding confirms the mainstream theory that Homo sapiens conquered the world "out of Africa."

Key success factors for Homo sapiens' dominance are related to bipedalism and brain size. Bipedalism—walking on two legs—

Our Animal Connection: What Sapiens Can Learn from Other Species (Second Edition)
Michael Hehenberger and Zhi Xia
Copyright © 2021 Jenny Stanford Publishing Pte. Ltd.
ISBN 978-981-4877-50-3 (Hardcover), 978-1-003-13072-7 (eBook)
www.jennystanford.com

raises the head, thereby permitting a greater field of vision with improved detection of distant dangers or resources. While walking upright, non-locomotory limbs (like hands) become free for other uses. The evolution of the opposable human thumb, and possibly also modifications in the ankle or foot that allow humans to walk on two legs, is still subject to discussion: walking may have been a by-product of busy hands.

Brain size increased from about 600 cubic centimeters to 1250 for Homo erectus and has now stabilized at around 1400. The human brain weighs about 1.5 kg and consumes about 20% of all energy provided by food.

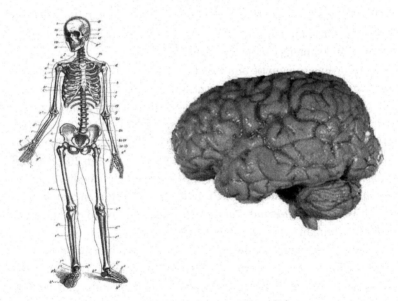

Figure 4.1 Bipedalism and brain size (source: Wikipedia).

Remembering the evolution of life on our planet, it is remarkable that the actions of Homo sapiens on the Earth's biosphere are quite recent but have already led to major changes such as integrity of the ozone layer, the proliferation of greenhouse gases, the conditions of productive soils, clean air and water, and other areas of significant (mostly negative) impact.

It is equally remarkable that our genetic makeup remains a reflection of the history of life on Earth.

To compare and connect sapiens to other living species, we need to introduce a few concepts from the world of medicine and physiology.

The human body is composed of many different types of cells that together create *tissues* and subsequently *organ systems.*

There are many different types of tissue, defined as cells with a specialized function. The study of tissues is called *histology* and typically requires a microscope. The body consists of four main types of tissues:

1. lining cells (epithelia) that line the cavities and surfaces of blood vessels and organs throughout the body;

2. connective tissue that consist of three main components: fibers (elastic and collagenous fibers), ground substance, and cells. In the central nervous system, connective tissue envelops the brain and spinal cord;

3. nervous tissue, included in the two parts of the central nervous system (CNS), the brain and spinal cord. Also included are the branching peripheral nerves of the peripheral nervous system (PNS) which regulate and control bodily functions and activity. Nervous tissue is composed of neurons, or nerve cells, which receive and transmit impulses, and glial cells which provide nutrients to the neurons;

4. muscle tissue that composes muscles in all animal bodies and gives rise to their ability to contract. The three types of muscle tissue are (1) skeletal or striated muscle, (2) smooth or non-striated muscle, and (3) cardiac muscle, sometimes referred to as semi-striated. Only skeletal/striated muscle contracts voluntarily, i.e., can be controlled by our brain. The smooth and cardiac muscle types are activated both through interaction of the central nervous system and/or via endocrine (hormonal) activation. Our cardiac muscle is "completely autonomous."

The human body comprises a head, neck, trunk (which includes the thorax and abdomen), arms and hands, legs and feet.

The study of the human body includes *physiology*, which focuses on the systems and organs of the human body and their functions. Many systems and mechanisms interact in order to maintain body temperature and safe levels of substances such as sugar and oxygen in the blood.

For a detailed list of human organs and their respective function we refer to the medical literature. It is worth, however, to point out a few facts that will be needed when exploring the sapiens connection to other species.

Musculoskeletal System

This system provides form, support, stability, and movement to the body. It is made up of the bones of the skeleton, muscles, cartilage, tendons, ligaments, joints, and other connective tissue. Its primary functions are to support the body, to enable motion, and to protect vital organs such as the heart, the lungs, liver and kidneys, and the brain. The skeletal portion of the system also provides a capacity to store calcium and phosphorus.

The yellow bone marrow has fatty connective tissue and can be used for energy in emergency situations, such as starvation. The red bone marrow is an important site for blood cell production, to replace existing cells that have died.

Ligaments connect the ends of bones together in order to form joints. They limit dislocation or prevent certain movements that may cause breaks. They can lengthen when under pressure and may be susceptible to break resulting in an unstable joint. Knees of athletes seem to be particularly exposed to ligament damage, definitely representing an area of human "weakness."

The physiology of human muscles is also worth a brief discussion, in view of what we have learned about ATP in a previous chapter: our body produces the majority of its ATP aerobically in the mitochondria without producing lactic acid or other byproducts causing fatigue or pain. When exercising our muscles, a second "nonstandard" method of ATP production can come into play. During activity that is higher in intensity, ATP production can switch to another (but less efficient) pathway, referred to as "anaerobic glycolysis." Anaerobic ATP production may have occurred during the early days of life on Earth, when there was no oxygen in the atmosphere. It produces ATP and allows near-maximal intensity exercise, but also produces significant amounts of lactic acid which limit the sustainability of such high intensity exercise to only a few minutes.

Vigorous physical activity increases the body's demand for oxygen. Oxygen consumption is dictated by the quantity of blood distributed by the heart, the ability of the lung to oxygenate the blood, and the working muscle's ability to take up the oxygen within that blood. Clearly, the oxygen carrying capacity of the blood, associated with red blood cell count, is important. Hypoxia, a condition where the body is deprived of adequate oxygen supply at the tissue level (e.g., at high altitude), reduces physical performance.

Digestive System

The human digestive system consists of the gastrointestinal tract and other organs of digestion such as the tongue, salivary glands, pancreas, liver, and gall bladder. The process of digestion has many stages, the first of which starts in the mouth. Digestion involves the breakdown of food into smaller and smaller components, until they can be absorbed and assimilated into the body.

The liver detoxifies food metabolites, synthesizes proteins, and produces biochemical compounds necessary for digestion. The liver is functioning as an accessory digestive gland that produces bile, an alkaline compound which helps the breakdown of fat. Bile is stored in a small pouch that sits just under the liver, named the gall bladder. Terminology related to the liver often starts in *hepat*, the Greek word for liver. Liver tissue consists mostly of "hepatocytes," which regulate a wide variety of high-volume biochemical reactions, including the synthesis and breakdown of both small and complex molecules.

More details will be discussed below in our microbiome chapter.

Respiratory System

The respiratory system of animals and plants is used for "gas exchange." In land animals, including humans, the linings of the lungs are forming the respiratory surface. As already discussed in a preceding chapter, gas exchange in the lungs of mammals occurs in millions of small air sacs called alveoli (as opposed to "atria" in birds). These air sacs communicate with the external environment via a system of airways, or hollow tubes, of which the largest is the trachea. As shown in Fig. 4.2, air is conducted from the environment into the alveoli (or atria) via conducting passages, by the process of breathing, facilitated by the muscles of respiration.

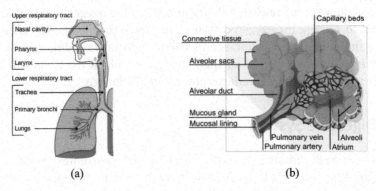

(a)   (b)

Figure 4.2 (a) Conducting passages of the human respiratory system. (b) An alveolus (Latin for "little cavity") is a microscopic structure found in the lung (source: Wikipedia).

Urinary System

The urinary (or renal) system consists of the kidneys, ureters, bladder, and the urethra, as shown in Fig. 4.3. The kidney's functional units are called nephrons. The renal system not only eliminates waste but regulates blood volume and blood pressure, manages electrolytes and metabolites, and regulates blood pH. Blood is supplied to the kidneys via renal arteries. Following blood filtration, it leaves the kidneys via the renal vein. Waste products (in the form of urine) exit the kidney via the ureters, tubes made of smooth muscle fibers. They move urine towards the urinary bladder, where it is stored and subsequently expelled from the body by urination.

The kidneys' filtration membranes clean the body's blood, making sure that water, small molecules, and ions can easily pass through. However, larger molecules such as proteins and blood cells are prevented from passing through the filtration membrane. The amount of filtrate produced every minute is called the glomerular filtration rate or GFR, reaching up to about 200 liters per day. Only about 1% of the filtrate (about 2 liters per day) becomes urine.

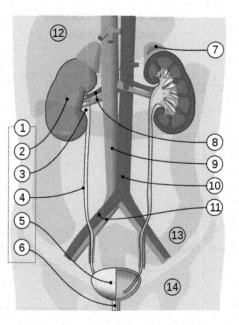

Figure 4.3 Urinary system with kidneys, *red* renal arteries, *blue* veins, ureters, and bladder (source: Wikipedia).

Endocrine System

The endocrine system is defined as the collection of glands that secrete hormones directly into the circulatory system to be carried towards target organs, resulting in "endocrine signaling." The major endocrine glands include the pineal gland, pituitary gland, pancreas, ovaries, testes, thyroid gland, parathyroid gland, and adrenal glands, as shown on Fig. 4.4.

The hypothalamus (see Fig. 4.5) is the neural control center for all endocrine systems. It links the nervous system to the endocrine system via the pituitary gland. Endocrinology is the branch of internal medicine that covers the endocrine system and its disorders.

The endocrine system is accompanied by the exocrine system, which secretes its hormones to the outside of the body using ducts.

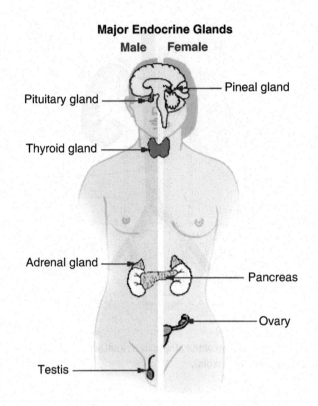

Figure 4.4 Major endocrine (male and female) glands (source: Wikipedia).

The endocrine system's effects are complex but rather slow as compared to responses of the nervous system. Endocrine signals may last from a few hours up to weeks. The nervous system sends information quickly and responses are short-lived.

Selected examples of important hormones are listed below. Obviously, their actions are highly relevant for our "body chemistry." Any imbalance or disrupture may have a major impact on our health and well-being.

Figure 4.6 shows a list of hormones related to the hypothalamus, thyroid, pineal and pituitary glands. The hypothalamus controls the anterior pituitary which then regulates various endocrine glands. Releasing hormones (or "releasing factors") are produced in hypothalamic nuclei, then often transported along axons to the posterior pituitary, where they are stored and released as needed.

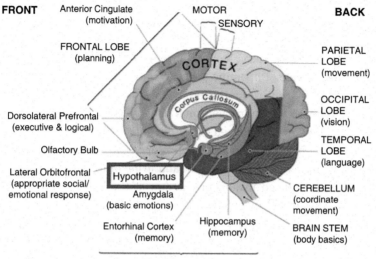

Figure 4.5 Human brain showing the location of hypothalamus (source: Wikipedia).

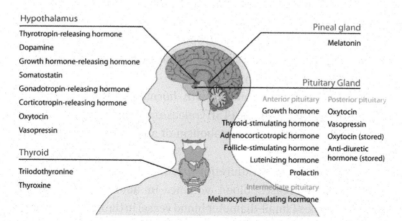

Figure 4.6 Anatomy and hormones associated with some important endocrine systems (source: Wikipedia).

Hypothalamus and Some Associated Hormones

- *Dopamine* inhibits prolactin, a hormone that enables milk production and acts as a regulator of the immune system. Dopamine also functions as a neurotransmitter, a chemical released by nerve cells (neurons) to send signals to other nerve cells. Dopamine pathways are known to play a major role in reward-motivated behavior. Many addictive drugs increase dopamine neuronal activity. Other brain dopamine pathways are involved in motor control. In general, dopamine functions as a local chemical messenger associated with increased sodium excretion and urine output in our kidneys, with reduced insulin production in the pancreas, and with reduced activity of lymphocytes in the immune system.

- *Growth hormone-releasing hormone* stimulates growth hormone (GH) release from the anterior pituitary. GH stimulates growth, cell reproduction, and cell regeneration. The name somatotropic hormone (STH) refers to the growth hormone produced naturally in animals (extracted from carcasses), while GH extracted from human cadavers is abbreviated hGH. Recombinant GH, i.e., formed by laboratory methods of genetic recombination (such as molecular cloning), is called somatropin and used as a prescription drug to treat children's growth disorders and adult growth hormone deficiency.

- *Somatostatin* inhibits growth hormone (GH) and thyroid-stimulating hormone (TSH) release from the anterior pituitary. TSH stimulates the metabolism of almost every tissue in the body.

- *Vasopressin* (anti-diuretic) promotes water reabsorption, which increases blood volume. In addition, it constricts arterioles (small-diameter blood vessel in the microcirculation that extends and branches out from an artery and leads to capillaries), leading to increased peripheral vascular resistance, thereby *raising* arterial blood pressure.

Pineal Gland

- *Melatonin* monitors the circadian rhythm, including induction of drowsiness and lowering of the core body temperature. Melatonin is also involved in blood pressure regulation. As a (non-prescription) medicine, it is used for the treatment of insomnia and—by frequent travelers—as "jet lag remedy."

Thyroid

- *Triiodothyronine and thyroxine* stimulate body oxygen and energy consumption, thereby increasing the basal metabolic rate. The thyroid hormones increase the basal metabolic rate, affect protein synthesis, help regulate long bone growth (synergy with growth hormone) and neural maturation. They also increase the body's sensitivity to catecholamines (such as adrenaline). The thyroid hormones are essential to proper development and differentiation of all cells of the human body. They regulate protein, fat, and carbohydrate metabolism. They also stimulate vitamin metabolism and are associated with heat generation. On the other hand, *thyronamines*, a family of metabolites of the thyroid hormones, are known to play an important role in the hibernation cycles of mammals. More research is needed to fully understand how they seem to inhibit neuronal activity.
- *Calcitonin* stimulates osteoblasts and thus bone construction; inhibits calcium release from bone, thereby reducing blood Ca^{2+}.

Stomach

- *Gastrin* secretes gastric acid.
- *Ghrelin* stimulates appetite.
- *Neuropeptide Y* increases food intake and decreases physical activity; can be associated with obesity.
- *Somatostatin* suppresses release of gastrin and other hormones; lowers rate of gastric emptying; reduces smooth muscle contractions and blood flow within the intestine.

- *Histamine* stimulates gastric acid secretion.
- *Endothelin* causes smooth muscle contraction of the stomach.

Duodenum (small intestine)

- *Secretin* secretes bicarbonate from liver, pancreas, and duodenal Brunner's glands; enhances effects of cholecystokinin; stops production of gastric juice.
- *Cholecystokinin* releases digestive enzymes from the pancreas; releases bile from the gall bladder; acts as hunger suppressant.

Liver

- *Insulin-like growth factor* (IGF) regulates cell growth and development; IGFs are proteins with high sequence similarity to insulin.
- *Angiotensinogen* and *angiotensin* cause vasoconstriction (regulating blood pressure); release aldosterone from adrenal cortex dipsogen. A dipsogen is an agent that causes thirst.
- *Thrombopoietin* stimulates megakaryocytes (large bone marrow cells responsible for the production of blood thrombocytes or "platelets," necessary for normal blood clotting).
- *Hepcidin* inhibits intestinal iron absorption and iron release by macrophages (a type of white blood cell that digests cellular debris, foreign substances, microbes, cancer cells, etc., that are not specific to healthy body cells, in a process called phagocytosis).

Pancreas

- *Insulin*, produced in beta (β) islet cells, reduces glucose in the blood. It also manages (in the liver) the intake of lipids and the synthesis of adipocytes (fat cells, primarily composed of adipose tissue, specializing in storing energy as fat).
- *Glucagon*, produced in alpha (α) islet cells, increases the blood glucose level via liver function.

- *Somatostatin*, produced in delta (δ) islet cells, inhibits the release of insulin, and inhibits release of glucagon. It suppresses the exocrine secretory action of the pancreas.

Kidney

- *Renin* activates the renin-angiotensin system (regulation of the plasma sodium concentration and arterial blood pressure).
- *Erythropoietin* (EPO) stimulates erythrocyte production. Red blood cells, also called erythrocytes, are the most common type of blood cell and our principal means of delivering oxygen (O_2) to the body tissues—via blood flow through the circulatory system.
- *Calcitriol* is an active form of vitamin D3 that increases absorption of calcium and phosphate from the gastrointestinal tract and kidneys and inhibits release of PTH, a hormone that is important in bone remodeling. PTH essentially increases blood calcium levels.
- *Thrombopoietin* stimulates megakaryocytes to produce platelets (see above: liver).

Please note that we just covered a part of the endocrine system. For instance, we left out sexual hormones involved in reproduction. In modern medicine, many bio-pharmaceutical treatments are designed to stimulate/repair hormone production. For example, diabetic patients have lost the full function of their pancreatic beta cells and must therefore provide the missing blood glucose reducing hormone by means of insulin injections. Similarly, patients suffering from insufficient red blood cell levels, have to inject EPO.

Cardiovascular System

The human cardiovascular system consists of heart, blood, and blood vessels. It includes the pulmonary circulation. Blood oxygenation happens through the lungs. As shown in Fig. 4.7, our hearts and lungs work together to provide oxygenated blood. An average adult body contains 5 to 6 liters of blood. Our blood represents about 7% of body weight. Blood consists of plasma, red blood cells, white blood cells, and platelets.

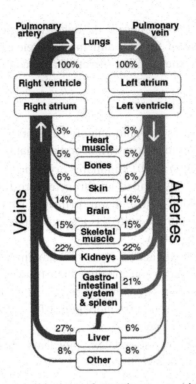

Figure 4.7 The heart and lungs work together to provide oxygenated blood (source: Wikipedia).

The cardiovascular system of humans is "closed," i.e., blood never leaves the network of blood vessels. However, oxygen and nutrients—acquired through the digestive system—diffuse across the blood vessel layers and enter interstitial fluid: oxygen and nutrients are carried to the target cells, and carbon dioxide and wastes are moved in the opposite direction. Figure 4.7 shows how cardiac output is delivered to various parts of the body.

The hemoglobin molecule plays a key role as the primary transporter of oxygen in mammals. In a healthy human, it carries 98.5% of the oxygen in the arterial blood. Only 1.5% of the oxygen is physically dissolved in the other blood liquids and not connected to hemoglobin.

Lymphatic System

The lymphatic system is another circulatory system and plays an important role as part of the *adaptive immune system*. The immune system protects the body against disease. To function properly, an immune system must detect a wide variety of agents, known as *pathogens*, from viruses to parasitic worms, and distinguish them from the organism's own healthy tissue. It is important to distinguish the *innate* immune system from the *adaptive* immune system.

The *innate immune system* comprises the cells and mechanisms that defend the host from infection by other organisms. It is the immune system we are born with. Innate immune systems provide immediate defense against infection, and include the following functions:

- Recruiting immune cells to sites of infection through the production of *chemical factors*, including specialized chemical mediators called *cytokines*
- Identification of bacteria, activation of cells, and clearance of antibody complexes or dead cells
- Use of specialized white blood cells to identify and remove foreign substances present in organs, tissues, blood and lymph
- Acting as a physical and chemical barrier to infectious agents

In evolutionary terms, the innate immune system has a longer history and it does not require any training or adaptation. One of its important functions is to be involved in the activation of the *adaptive immune system* through a process known as antigen presentation.

The *adaptive immune system*, also known as the *acquired* immune system, is composed of highly specialized, systemic cells and processes that eliminate pathogens or prevent their growth. Adaptive immunity is triggered by an initial response to a specific pathogen. When that happens, immunological memory is created. In subsequent encounters that pathogen will be recognized and will result in an enhanced response. This process of acquired immunity is the basis of vaccination.

The lymphatic vessels carry a clear fluid called *lymph*, very similar to blood plasma: lymph contains lymphocytes and they are concentrated in the lymph nodes. The spleen and the thymus are also lymphoid organs of the immune system. Another interesting role is played by the tonsils: they are both lymphoid organs and are associated with the digestive system.

Lymphocyte tissues are produced in the thymus and the bone marrow.

The name "T cells" or "T lymphocytes" is based on the fact that they mature in the thymus from thymocytes. A T cell is a subtype of white blood cell that plays a central role in cell-mediated immunity. T cells can be distinguished from other lymphocytes, such as B cells, by the presence of a T-cell receptor on the cell surface. The number of T cells is kept in balance by the process of *apoptosis*, a form of programmed cell death.

"B cells," on the other hand, mature in the bone marrow, which is at the core of most bones. Their function in the adaptive immune system is to secrete antibodies. B cells express B cell receptors (BCRs) on their cell membrane. In turn, BCRs allow the B cell to bind to a specific antigen, thereby initiating an antibody response.

The lymphatic system plays a major role as the primary site for cells relating to our adaptive immune system. Cells in the lymphatic system react to *antigens* presented or found by those cells. After recognition of an antigen, an immunological cascade begins with the activation and recruitment of more and more cells. Subsequently, antibodies and cytokines are produced and other immunological cells (such as *macrophages*) are recruited.

Nervous System

The human brain is the central organ of the human nervous system. Along with the spinal cord, it makes up the central nervous system (CNS).

Humans owe their success mostly to the brain. It consists of the cerebrum, the brainstem and the cerebellum. The brain controls most of the activities of the body, processing, integrating, and coordinating the information it receives from the sense organs, and making decisions as to the instructions sent to the rest of the body. The brain is contained in, and protected by, the skull bones of the head. The *cerebrum* is divided into two cerebral hemispheres. Each hemisphere is normally divided into four lobes: the frontal, temporal, parietal, and occipital lobes (see Fig. 3.5).

The frontal lobe is associated with executive functions, including self-control, planning, reasoning, and abstract thought, while the occipital lobe is dedicated to vision. Cortical areas within each lobe are associated with specific functions, such as the sensory, motor, and association regions.

Although the left and right hemispheres are broadly similar in shape, some functions are associated with one side, such as language in the left, and visual-spatial ability in the right. The hemispheres are connected by nerve tracts, the largest being the *corpus callosum*.

There are more than 86 billion neurons in the brain. All brain activity is enabled by the interconnections ("synapses") of neurons and their release of neurotransmitters in response to nerve impulses. Neurons form elaborate "neural networks" of pathways and circuits that are developed as babies are growing up and learn.

In newborn human babies, neurons are quite unconnected to each other. During the first three years, the brain cells become more and more interconnected. After 3 years, the connections are pruned back, with the remaining connections becoming stronger and adapting themselves to the particular existing environment. For the brains of babies to develop properly, they need extensive stimulation during their first 3 years. The stimulation should include cognitive learning aspects as well as emotional care. Brain development then continues and adolescence is another period of significant cognitive change.

Although the human brain is protected by the skull, suspended in cerebrospinal fluid, and isolated from the bloodstream by the blood–brain barrier, it still quite susceptible to damage, disease, and infection. Humans are not very well protected against brain damage that can be caused by trauma, or by a loss of blood supply known as a stroke. The human brain is further susceptible to degenerative disorders, such as Parkinson's disease, dementias including Alzheimer's disease, and multiple sclerosis. Among brain dysfunctions we can also mention psychiatric conditions such as schizophrenia and clinical depression.

The spinal cord extends down all the way between the first and second lumbar vertebrae. It supports the transmission of nerve signals from the motor cortex to the body, and from the fibers of the sensory neurons to the sensory cortex. The spinal cord further coordinates many of the body's reflexes.

A *nerve* is defined as a cable-like, enclosed, bundle of *axons* (nerve fibers, projections of neurons) in the peripheral nervous system. It typically conducts electrical impulses known as action potentials, away from the nerve cell body. Axons transmit information to different neurons, muscles, and glands by having electrochemical impulses travel from peripheral organs to the cell body. This information is then propagated from the cell body to the spinal cord. In addition to neurons, nerves also include non-neuronal cells that coat the axons in myelin. There are two types of axons in the nervous system, namely myelinated and unmyelinated axons. *Myelin* is a layer of a fatty insulating substance and plays an important role in brain health.

Axons make contact with other cells at junctions called *synapses*. At a synapse, the membrane of the axon closely adjoins the membrane of the target cell, and special molecular structures transmit electrical or electrochemical signals across the gap.[a] A bundle of axons forms a *nerve tract* in the central nervous system, and a *fascicle* in the peripheral nervous system. The largest white matter tract in the brain is the above-mentioned corpus callosum, comprising some 20 million axons.

Neurotransmitters are chemical messengers that transmit signals across a synapse, such as a neuromuscular junction, from one neuron to another "target" neuron, muscle cell, or gland cell of the endocrine system. Examples of neurotransmitters are listed below:

- *Glutamate* is used at the great majority of fast synapses in the brain and spinal cord. Excessive glutamate release can overstimulate the brain and lead to *excitotoxicity* causing cell death resulting in seizures or strokes. Excitotoxicity has been implicated in certain chronic diseases, including ischemic stroke, epilepsy, amyotrophic lateral sclerosis, Alzheimer's disease, Huntington disease, and Parkinson's disease.[46]
- *GABA* (γ-aminobutyric acid) is used at the great majority of fast *inhibitory* synapses in virtually every part of the brain. Many sedative/tranquilizing drugs act by enhancing the effects of GABA.
- *Acetylcholine* was the first neurotransmitter discovered in the peripheral and central nervous systems. It is distinguished

---

[a]We will return to molecular details when discussing the sea slug in Chapter 6A.

as the transmitter at the neuromuscular junction connecting motor nerves to muscles.

- *Dopamine* is playing a number of important functions in the brain. Dopamine regulates motor behavior and is related to pleasure as well as motivation. It plays a critical role in the reward system.

- *Serotonin* is a monoamine neurotransmitter. It is an interesting fact that 90% of serotonin is produced by, and found in the intestine. Serotonin also plays important roles in central nervous system neurons, regulating appetite, sleep, memory and learning, temperature, mood, behavior, muscle contraction, and function of the cardiovascular system and endocrine system. It may also play a role in depression.

- *Norepinephrine* is synthesized (from tyrosine) in the central nervous system and modulates the responses of the autonomic nervous system, the sleep patterns, focus and alertness.

- *Epinephrine* is also synthesized from tyrosine and released in the adrenal glands and the brainstem. It plays a role in sleep, influences the ability to be and stay alert, and the "fight-or-flight" response.

- *Histamine* works with the central nervous system (CNS), specifically the hypothalamus and CNS mast cells. Mast cells are a type of white blood cell, best known for their role in allergy and in important protective roles such as wound healing, angiogenesis (formation of new blood vessels), immune tolerance, and defense against pathogens.

Sensory Nervous System

When exploring our animal connection, one of the most important areas is the part of the human nervous system responsible for the processing of sensory information. Each sensory system consists of sensory neurons, neural pathways, and the particular parts of the brain involved in sensory perception. Commonly recognized sensory systems are those for the five senses of vision, hearing, touch, taste, and smell. Beyond those conventional and broadly recognized senses, there are other "sensory modalities," including temperature, pain, balance, and vibration. Figure 4.8A shows at high levels of complexity how sensory signals are communicated inside the spinal cord.

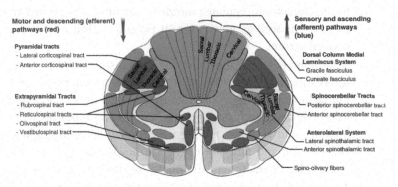

Figure 4.8A Motor pathways and sensory pathways inside the spinal cord (source: Wikipedia).

The anterolateral system (or spinothalamic tract) is a sensory pathway from the skin to the thalamus, a part of the brain shown in Fig. 4.8B.

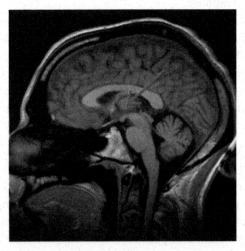

Figure 4.8B Thalamus (marked) from MRI cross-section (source: Wikipedia).

The thalamus performs several important brain functions, including the relaying of sensory signals to the cerebral cortex, and the regulation of consciousness, sleep, and alertness.

The spinothalamic tract consists of two pathways, namely anterior and lateral. The anterior spinothalamic tract carries information about *crude touch and firm pressure*. The lateral spinothalamic tract conveys *pain and temperature*. Similarly, the dorsal column-medial lemniscus tract conveys sensory information from the periphery to the conscious level at the brain's cerebral cortex.

The **skin** is also playing an important role in our sensory system: In hairless skin, there are four principal types of mechanoreceptors, each shaped according to its function. The tactile corpuscles (or *Meissner corpuscles*) respond to light touch and adapt rapidly to changes in texture (vibrations around 50 Hz). The bulbous corpuscles (or *Ruffini endings*) detect tension deep in the skin. The *Merkel nerve endings* detect sustained pressure. The lamellar corpuscles (or *Pacinian corpuscles*) in the skin detect rapid vibrations (of about 200–300 Hz).

Our brain has an amazing ability to process and integrate what our senses transmit from the physical world around us, and to link the information with past experiences, thereby creating a unique and personalized perception of the world around us. As explained by David Eagleman in his popular book *The Brain*,[47] we are somehow able to create a single, unified picture of the world despite the fact that the electrochemical signals associated with vision, hearing, touch, taste and smell are processed in different regions of the brain.

Human Eye (Vision)

Sight or vision is the capability of the eye(s) to focus and detect images of visible light. As defined by Eric Kandel,[48] for humans "vision is the process of discovering from images *what* is present in the visual world and *where* it is." What we normally see is a three-dimensional, moving colored image. Rod and cone cells in the retina do the job. *Rods* are very sensitive, even working in low light, but only perceiving shades of grey, from black to white. *Cones,* on the other hand, require daylight and are less sensitive, but capable of color differentiation. A third type of human eye cells are non-image-forming photosensitive ganglion cells in the retina. Those cells react to light signals that trigger (i) adjustment of the size of the pupil and (ii) regulation and suppression of the hormone melatonin. Melatonin is known to be associated with our "body clock." Figure 4.9 shows the most important components of the eye.

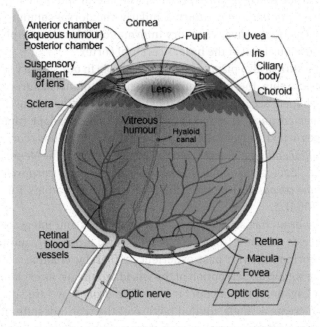

Figure 4.9   Horizontal section through human eye (source: Wikipedia).

Light energy enters the eye through the cornea, through the pupil and then through the lens. The ciliary muscle controls the lens shape, to make sure it will be changed for near focus (accommodation). When light photons are hitting the retina's photoreceptor cones, they are converted into electrical signals that are then transmitted to the brain via the optic nerve. There are about 100 million sensory cells in the retina of a human healthy eye.

The visual system in the human brain is quite slow and requires a significant portion (about one-third!) of overall brain capacity. Having two eyes allows the brain to determine the depth and distance of an object, called stereovision, and to create a three-dimensional image. To stimulate stereovision, both eyes must point accurately to the object and coordinate the two respective retinas. Kandel's "what" and "where" questions are processed by the brain along parallel paths. Upon integration of these paths, the present image is further compared to images encountered in the past. In particular, most of the brain's computational power is devoted to face recognition, an essential requirement for sapiens to function socially.

For the human eye, the visible portion of light is electromagnetic radiation with wavelengths in the range of 400–700 nm, falling between the "longer" infrared (IR) wavelengths and the "shorter" ultraviolet (UV) wavelengths. As seen on Fig. 4.10, the visible light portion accessible to the human eye represents only a small portion of the full electromagnetic spectrum.

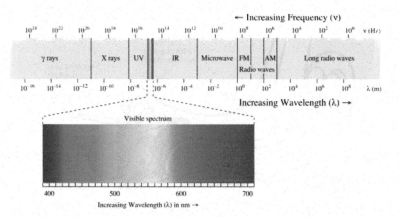

Figure 4.10 Electromagnetic spectrum showing the "visible portion" (source: Wikipedia).

Note that animals with eyes that do not require lenses (e.g., insects and shrimp) are able to detect ultraviolet, in much the same chemical way that humans detect visible light. While humans have three types of cone cells, birds possess a forth type that extends their vision beyond the human range into the UV area.

Human Ear (Hearing)

Hearing is the sense of sound perception. Sound vibrations are mechanically conducted from the eardrum through a series of tiny bones to hair-like fibers in the inner ear. These fibers are capable of performing mechanical motion within a range of about 20 to 20,000 Hz. However, our perception of high frequencies declines with age. The inability to hear is called deafness.

Sound waves are reflected and attenuated when they hit the cartilage ("*pinna*") surrounding the ear canal. The resulting anatomical changes provide information that will help the brain determine the direction from which the sounds came. Sound waves travel through the ear canal, hit the eardrum and move across the air-filled middle ear cavity via a series of delicate bones or "ossicles": as illustrated in Fig. 4.11, the three ossicles malleus (hammer), incus (anvil), and stapes (stirrup) convert the lower-pressure eardrum sound vibrations into higher-pressure sound vibrations at another, smaller membrane called the oval window or vestibular window.

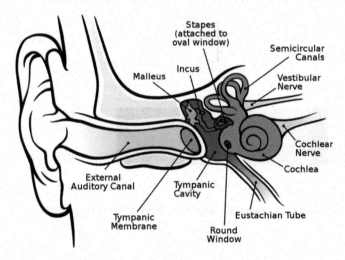

Figure 4.11 Anatomy of the human ear (source: Wikipedia).

The hair cells of the inner ear ("organ of Corti") convert sound waves into nerve impulses that can be processed by the brain. Compared to our vision, hearing requires a lot less of the brain's processing power: there are only about 16,000 hair cells that detect sound and movement, and convert this input into electrical signals. Further processing leads to a wide spectrum of auditory reactions and sensations, including the exceptional emotional impact (joy, sadness, etc.) created by music.

The auditory system is delicate and sensitive and must be fully functional to process and understand sound input. Impaired hearing has the potential to adversely impact an individual's ability to communicate, learn, and effectively complete routine tasks. In children, early diagnosis and treatment of an impaired auditory system function is critical. The brain's frontotemporal system underlying auditory perception permits humans distinction of sounds as speech, music, or noise.

Human Touch

Touch, or somatosensation, also called tactition or mechanoreception, is a perception resulting from activation of neural receptors, generally in the skin, including hair follicles, but also in the tongue, throat, and mucosa. Numerous pressure receptors respond to variations in pressure and are able to detect firm, "brushing," and sustained forms of pressure. The touch sense of *itching* caused by insect bites or allergies involves special itch-specific neurons in the skin and spinal cord. Loss or impairment of the ability to feel anything touched are referred to as tactile anesthesia. Paresthesia is a sensation of tingling, pricking, or numbness of the skin that may result from (permanent or temporary) nerve damage.

Recent brain science studies have revealed[49] that touch can add an additional dimension to visual perception: regardless of perception by either eye or hand, the texture of an object is activating a special region of the brain. As pointed out in Kandel's book *Art and Brain Science*, vision and touch are further capable of recruiting the emotional systems of the brain, including amygdala,[b] the hypothalamus[c] (see Fig. 4.5), and the dopaminergic modulatory system.

Human Taste

Taste, or gustation, refers to the capability to detect the taste of substances, including food, certain minerals, poisons, etc. The sense of taste is often confused with the "sense" of *flavor*, which is a combination of taste and smell perception. Flavor depends on odor, texture, and temperature as well as on taste. Humans receive tastes through sensory organs called taste buds, concentrated on the

---

[b]Among other functions, the amygdala orchestrates emotions (both positive and negative).
[c]Among other functions, the hypothalamus makes us "feel emotion."

upper surface of the tongue. There are five basic tastes: sweet, bitter, sour, salty, and umami (Japanese for "savory taste"). Umami taste receptors respond to glutamate, a generic name for flavor-enhancing compounds based on glutamic acid and its salts (glutamates). Glutamic acid is obtained by treating wheat gluten with sulfuric acid. Gluten is a composite of storage proteins found in wheat, barley, rye, oats, and related species. Glutamic acid and glutamates are natural constituent of many fermented or aged foods, including soy sauce, fermented bean paste, cheese, and yeast extracts. Glutamates are ubiquitous in biological life.

The umbrella term "gluten-related disorders" is used for all diseases triggered by gluten. They include celiac disease (CD), non-celiac gluten sensitivity (NCGS), wheat allergy, etc.

CD is an autoimmune disorder affecting ~1% of the population. It is caused by the ingestion of wheat, barley, rye, and derivatives and causes health issues in genetically predisposed people of all ages. "Classic" celiac disease symptoms include gastrointestinal issues such as chronic diarrhea and abdominal distention, loss of appetite, and impaired growth. More seriously, CD with "non-classic symptoms" occurs in children over 2 years of age, adolescents, and adults. In addition to intestinal symptoms, its non-intestinal manifestations may involve any organ of the body. Both CD and non-celiac gluten sensitivity (NCGS, causing gastrointestinal symptoms but also headache and chronic fatigue), can be "cured" by switching to a gluten-free diet.

Human Smell

Smell or olfaction is a "chemical" sense, just as "taste." In 1991, Richard Axel and Linda Buck[50] discovered that hundreds of genes in our DNA are coding for the odorant sensors located in the olfactory sensory neurons in our noses. When an odorant attaches itself to the receptor, it triggers a protein change and an associated electric signal to be sent to the brain. Smells are composed of a large number of different substances and we interpret the varying signals from our receptors as specific scents. Odor molecules possess a variety of features and, thus, excite specific receptors more or less strongly. This combination of excitatory signals from different receptors makes up what we perceive as the molecule's "smell." In the brain, olfaction is processed by the olfactory system. Olfactory receptor neurons in the **nose** differ from most other neurons in that they die and regenerate on a regular basis. Figure 4.12 shows the peripheral olfactory system, which consists mainly of the nostrils, ethmoid bone, nasal cavity, and the olfactory epithelium, characterized by layers of thin tissue (covered in mucus) that line the nasal cavity. The layers of epithelial tissue include the mucous membranes, olfactory glands, olfactory neurons, and nerve fibers of the olfactory nerves.

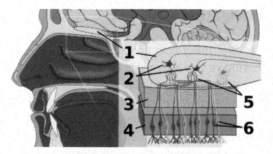

Figure 4.12 Human olfactory system: (1) olfactory bulb, (2) mitral cells, (3) bone, (4) nasal epithelium, (5) glomerulus, (6) olfactory receptor cells (source: Wikipedia).

The *glomerulus* is a spherical structure located in the olfactory bulb of the brain. Inside the glomerulus, synapses form between the terminals of the olfactory nerve and the dendrites of receptor neurons.

Odor molecules can enter the peripheral pathway and reach the nasal cavity in one of two ways: (i) through the nostrils when inhaling (olfaction) or (ii) through the throat when the tongue pushes air to the back of the nasal cavity while chewing or swallowing (retronasal olfaction). Inside the nasal cavity, mucus lining the walls of the cavity dissolves odor molecules. Mucus also covers the olfactory epithelium. It contains mucous membranes that produce and store mucus, and olfactory glands that secrete metabolic enzymes found in the mucus.

Some neurons in the nose are specialized to detect pheromones. A *pheromone* is defined as a chemical factor that triggers a social response in members of the same species. Normally, pheromones are excreted. They are capable of acting like hormones outside the body of the secreting individual, to impact the behavior of the receiving individuals. Different types of pheromones include alarm pheromones, food trail pheromones, sex pheromones, and many others that affect behavior or physiology. The use of pheromones extends across all living species, from basic unicellular prokaryotes to complex multicellular eukaryotes. In particular, their use among insects has been well documented.

Finally, the inability to smell is called *anosmia*.

Spatial Navigation

The ability to find one's way under difficult conditions such as in darkness, in a fog, or in a blizzard is sometimes called a "sixth sense" and varies greatly between individuals.

The ability to navigate is now known to depend on specialized neurons in the brain's hippocampus area (see Fig. 4.5). Their detailed description was awarded by the 2014 Nobel Prize for Medicine and Physiology[51] to John O'Keefe, May-Britt Moser, and Edvard I. Moser. They received the award for their discoveries of cells that constitute a positioning system in the brain. In 1971, John O'Keeffe found that a type of nerve cell in the brain's hippocampus area was always activated when a rat was at a certain place in a room. Other nerve cells were activated when the rat was at other places. He concluded that these "place cells" formed a map of the room. In 2005, May-Britt and Edvard Moser discovered another key component of the brain's positioning system, namely "grid cells" that generate a coordinate system and allow for precise positioning and pathfinding. Their subsequent research showed how, together, place and grid cells make it possible to determine position and to navigate.

The "spatial navigation" problem has occupied philosophers and scientists for centuries: How does the brain create a map of the space surrounding us and how can we navigate our way through a complex environment? How do we experience our environment? The sense of place and the ability to navigate are fundamental to our existence. The sense of place gives a perception of position in the environment. During navigation, it is interlinked with a sense of distance that is based on motion and knowledge of previous positions.

Recent investigations with brain imaging techniques, as well as studies of patients undergoing neurosurgery, have provided evidence that place and grid cells exist also in humans. In patients with Alzheimer's disease, the hippocampus is frequently affected at an early stage, and these individuals often lose their way and cannot recognize the environment. Knowledge about the brain's positioning system may, therefore, help us understand the mechanism which may be responsible for the devastating spatial memory loss that affects people with this disease.

The discovery of the brain's positioning system represents a paradigm shift in our understanding of how ensembles of specialized cells work together to execute higher cognitive functions. It has opened new avenues for the understanding of other cognitive processes, such as memory, thinking, and planning.

What is Pain?

A discussion of the human sensory system cannot be complete without covering pain, a hard-to-define and quite subjective "unpleasant sensory and emotional experience associated with actual or potential tissue damage, or described in terms of such damage."[52]

Nociception (derived from the Latin verb *nocere*, which means "to harm") distinguishes the physiological process related to pain from the "subjective" experience of pain. In nociception, intense *chemical, mechanical,* or *thermal* stimulation of sensory nerve cells called nociceptors produces a "noxious" signal that travels along nerve fibers via the spinal cord to the brain. Nociceptors are nerve endings that detect such stimuli. They can be found in the skin, on internal body surfaces and in some internal organs. The concentration of nociceptors varies throughout the body. The categorization of nociceptors is based on the axons that travel from the receptors to the spinal cord or brain. Nociceptors require a minimum intensity of stimulation before they trigger a signal. Once this threshold is reached, a signal is passed along the axon of the neuron into the spinal cord.

On a mechanistic level, *transient receptor potential* (TRP) channels are related to action potentials in nerves that can be differentiated as follows:

1. chemical TRP channels that act like taste buds, signaling if their receptors bond to certain elements/chemicals (e.g., iodine in a cut of our skin);
2. mechanical TRP channels react to depression of their cells (like touch, but caused by crushing, tearing, shearing, etc);
3. thermal TRP channels change shape at temperatures outside the moderate range of 24–28°Celsius (cold or hot).

In addition to *nociceptive* pain, there are two more categories of pain, namely

- *inflammatory* pain, associated with tissue damage and the infiltration of immune cells, and
- *pathological* pain, caused by damage to, or abnormal function of, the nervous system.

Pain is usually transitory, lasting only until the noxious stimulus is removed or the damage has been dealt with, usually via healing. However, some painful conditions, such as rheumatoid arthritis, peripheral neuropathy, cancer, and *idiopathic* pain (conditions where the cause may not be readily apparent or characterized) may persist for years. Pain that lasts a long time is called *chronic* or persistent, and pain that resolves quickly is called *acute*.

Pain is part of the body's defense system, producing a reflexive retraction from the painful stimulus, along with protection of the affected body part while it heals. However, Homo sapiens is a very complicated creature, making it very difficult to understand the various sources of pain and to manage the symptoms.

An *analgesic* or painkiller is any member of the group of drugs used to achieve analgesia, i.e., relief from pain. Analgesic drugs act in various ways on the peripheral and central nervous systems. They are distinct from *anesthetics*, which temporarily affect, and in some instances completely eliminate, sensation.

The most popular and widespread drugs used to deal with pain are nonsteroidal anti-inflammatory drugs (NSAIDs). They reduce pain, decrease fever, prevent blood clots and, in higher doses, even decrease inflammation. Side effects depend on the specific drug, but largely include an increased risk of gastrointestinal ulcers and bleeds, heart attack, and kidney disease.

NSAIDs work by inhibiting the activity of *cyclooxygenase enzymes* (COX-1 and/or COX-2). In cells, these enzymes are involved in the synthesis of key biological mediators, namely *prostaglandins*, which are involved in inflammation, and *thromboxanes*, which are involved in blood clotting. The most prominent NSAIDs are aspirin, ibuprofen and naproxen, all available over the counter in most countries. Paracetamol (acetaminophen) is generally not considered an NSAID because it has only little anti-inflammatory activity. It treats pain mainly by blocking COX-2 mostly in the central nervous system, but not much in the rest of the body.

Another important category of pain medicines are the opioids. They are limited to the natural alkaloids found in the resin of the *opium poppy* although some include semi-synthetic derivatives. An important semi-synthetic opioid that is synthesized from codeine, one of the opioid alkaloids found in the opium poppy, is *hydrocodone*. It is a narcotic analgesic used orally for relief of moderate to severe pain, but also commonly taken in liquid form as an antitussive/ cough suppressant. Opioids act by binding to opioid receptors. They are found principally in the central and peripheral nervous system and the gastrointestinal tract. Opioid receptors mediate both the psychoactive and the somatic (associated with voluntary movements of the body) effects of opioids. Medically, they are primarily used for pain relief, including anesthesia. Other medical uses include suppression of diarrhea, suppressing cough, and suppressing opioid induced constipation. However, opioids are also frequently used non-medically for their *euphoric* effects or to prevent withdrawal. Side effects of opioids may include itchiness, sedation, nausea, respiratory depression, constipation, and euphoria. Tolerance and dependence will develop with continuous use, requiring increasing doses and leading to a withdrawal syndrome upon abrupt discontinuation. The euphoria attracts recreational use, and frequent, escalating recreational use of opioids typically results in addiction. Fatal consequences can be caused by an overdose or by concurrent use of opioids with other depressant drugs, resulting frequently in death from respiratory depression. This risk for addiction and fatal overdoses dictates that opioids are mostly controlled substances.

Morphine is a pain medication of the opiate variety which is found naturally in a number of plants and animals. It acts directly on the central nervous system to decrease the feeling of pain. It can be taken for both acute pain and chronic pain and is frequently used for pain from myocardial infarction, during labor, and by cancer patients. Long-acting formulations also exist. Morphine is the gold standard to which all *narcotics* are compared. Potentially serious side effects include a decreased respiratory function and low blood pressure. Morphine has a high potential for addiction and abuse.

Medical *cannabis* (or medical marijuana) refers to the use of cannabis (used for hemp fiber, for hemp oils, for medicinal purposes, and as a recreational drug) to reduce nausea and vomiting during chemotherapy, to improve appetite in people with HIV/AIDS, and to treat *chronic pain* and muscle spasms.

Cocaine, also known as coke, is a strong stimulant mostly used as a recreational drug. It has a small number of accepted medical uses such as numbing and decreasing bleeding during nasal surgery. Cocaine is addictive due to its effect on the reward pathway in the brain. It acts by inhibiting the reuptake of serotonin and dopamine, resulting in greater concentrations of these neurotransmitters in the brain. The resulting effect is loss of contact with reality along with an intense feeling of happiness. Cocaine is a naturally occurring substance found in the coca plant which is mostly grown in South America.

Because of their addiction risk, all opioids are usually reserved for moderate to severe pain.

Another way to manage pain is via "physical medicine and rehabilitation," by employing diverse techniques such as thermal agents and electrotherapy, as well as therapeutic exercise and behavioral therapy, alone or in tandem with drug therapy.

Acupuncture is believed by its followers to restore the energy balance in the body, thereby reducing pain signals through production of *endorphins* that are known to be the natural painkillers. Clinical studies suggest that acupuncture can reduce joint pain and so the therapy can be effective in reducing pain caused by knee osteoarthritis.

Interventional procedures—typically used for chronic back pain—include epidural steroid injections, spinal cord stimulators, and intrathecal drug delivery system implants. Pulsed radiofrequency, neuromodulation, direct introduction of medication, and nerve ablation may be used to target either the tissue structures and organ/systems responsible for persistent nociception or the nociceptors from the structures implicated as the source of chronic pain.

Sapiens Challenges and Weaknesses

Having covered some of the basic facts of human anatomy and physiology, let us now pause and discuss our challenges and weaknesses. Homo sapiens may be wise but we are also quite fragile, partly because we have learned to rely so much on our superior brain.

We can run and have surprisingly good endurance, but many animals can beat our speed. We can jump, but to many animals our leaps look quite ordinary. The many advantages of our upright walk are partly compensated by lower back pain, aching hips and knees, and by ligaments that cannot support extreme movements. That applies, e.g., to the ACL (anterior cruciate ligament) of our knees. It is frequently torn whenever athletes compete with each other. Another weak link is the ulnar collateral ligament (UCL) of our elbow which is frequently overused by (baseball) athletes who throw "fastballs." To repair a torn UCL's, the surgeon often replaces the elbow tendon with another tendon of the patient's body.

Above all, human females pay a very high price for sapiens' bipedal movement by having to endure an often difficult and painful childbirth: due to the anatomy of human musculoskeletal structure, giving birth to a child is frequently a traumatic, dangerous, and potentially damaging experience. Muscles can stretch and tear irreparably, the baby can become stuck or come out the wrong way, and in best case it is still a painful, although often emotionally rewarding, experience.

When sapiens evolved to walk upright, the shape of the pelvis changed and the birth canal became narrow. At the same time, as the human use of tools increased and social interaction became more frequent, sapiens' intelligence gradually increased, resulting in larger and larger brains. Evolution solved the problem by giving babies the ability to compress the skull as they pass through the birth canal. However, a mature brain would still be too large, so nature starts out with a minimal brain size at birth and then grows and develops the mature brain time after birth. The baby has then to learn how to control the muscle system and to produce more motor-related brain functions. Therefore, people walk only about a year after birth, which is basically the result of balance between pelvic reduction and brain volume increase in evolution.

Our senses, as described above, work fairly well and our brains are doing a fine job in processing and merging the various inputs, but we typically turn to artificial enhancements to sharpen our capabilities: we use microscopes and binoculars to extend our vision and even invent new ways to "see" the invisible by instruments such as the scanning tunneling microscope, a new visual concept that is based on the principle that the structure of a surface can be studied using a stylus that scans the surface at a fixed distance from it.

The human brain is our greatest asset. However, it can also malfunction and cause problems when the hormonal chemistry is out of order, or when neurons and synapses misbehave.

A simple way of describing our brain in evolutionary terms would be to divide it into reptilian functions, the limbic or mammalian brain, and the more recently evolved neocortex. In this oversimplified model, "reptilian" functions include instinctual behaviors involved in aggression, dominance, territoriality, and ritual displays. All motor functions and "fight-or-flight" instincts would belong.

The "limbic" brain functions would then include what we call "emotions," and the neocortex would enable our rational thinking.

Our minds are motivated by emotional systems such as fear, love, rage and grief. If the emotions are too strong, they can overpower rationality and lead to neuroses and maladaptive behavior. When discussing science versus religion,[53] one could place science squarely into the neocortex and religion at least partly into the limbic brain area. There are many forms of human emotion, in particular human suffering and grief, that may be better managed by religion than by science, i.e., purely rational thinking. Most humans therefore depend on religious rituals to help them get through troubled times.

When comparing humans and animals, a fundamental difference is that people can rely on reason to control their own behavior, and animals only rely on instinct. When relying on religion instead of scientific, rational thinking, humans seem to take a step back to simple, mammalian behavior and are then rewarded by religion to receive help in managing their emotions. However, it would be wrong to assume that science and religion are incompatible: while some prominent scientists deny the existence of God,[54] a number of outstanding scientists[55] are deeply religious and defend their belief in a superhuman power.

Being social has contributed a lot to human success. However, there is also a negative aspect to the combined effect of emotions and group dynamic on human behavior. Emotions can get out of control, overwhelm rational thinking and lead to destructive behavior, in particular if enhanced by group pressure. As social animals, humans behave differently when connected with like-minded people in groups. People may even commit violent actions they would avoid individually and may often regret afterwards.

Humans are the most destructive species. The more power we gain by using our superior brains, the higher the destructive risks. Humans have an urge to understand what they experience around them, to improve the tools they are using to gain a competitive advantage. The urge to understand, to dig ever deeper, has led to amazing progress in science and technology. The human urge to compete has driven sapiens to both work harder, but also to fight fellow humans, and to start wars. The combination of our basic urges to "understand" and to "compete," has led to the development of increasingly powerful weapons, weapons that in the hand of the "wrong people" could easily destroy our planet. Fortunately, the limits of science can be protected by ethics, a very positive human achievement that should never be forgotten when scientific breakthroughs are made. Also, the competitive spirit can be channeled into more harmless activities such as Sports. The Greeks who built one of the first advanced civilizations, already realized this dilemma and started the first Olympic Games dedicated to sports competitions instead of serious war games. In modern times, the wise men responsible for initiatives such as the Human Genome Project have added bioethics to their scientific objectives, and peace-loving politicians and educators are making sure that human competitive emotions and misdirected nationalistic movements are channeled into football or basketball matches instead of wars.

The recent inventions of computers—clearly based on and supporting scientific thinking—could possibly both reduce or increase those human brain-related challenges. As modern computer software is adding *artificial intelligence* functions to complement basic arithmetic capabilities, computers may become more human-like and therefore graduate from "decision support" to the rather controversial area of "decision making." According to Kurzweil,[56] nonbiological intelligences will claim to have consciousness and "the full range of emotional and spiritual experiences that humans claim to have." Furthermore, he asserts that the risk of having computers take over, after having surpassed the limits of human intelligence (Kurzweil's definition of "singularity"), can be mitigated by "increasing the values of liberty, tolerance, and respect for knowledge and diversity in society," because "the nonbiological intelligence will be embedded in our society and will reflect our values."

What is it That Kills Homo Sapiens?

Statistically,[57] the top ten leading causes of human mortality, measured by deaths per 100,000 population, are

1. Ischemic heart disease (119): IHD or coronary artery disease (CAD) refers to a group of diseases including angina, myocardial infarction, and sudden cardiac death.
2. Stroke (85): Stroke is a medical condition in which poor blood flow to the brain results in cell death. It can be caused by lack of blood flow (ischemic), or by hemorrhaging, due to bleeding. Signs and symptoms of a stroke may include an inability to move or feel on one side of the body, problems understanding or speaking, or loss of vision to one side.
3. Lower respiratory tract infections (43): LRTI is often used as a synonym for pneumonia but can also be applied to other types of infection including acute bronchitis, the bacterial or viral infection of the larger airways. Symptoms include shortness of breath, weakness, fever, coughing and fatigue.
4. Chronic obstructive pulmonary disease (43): COPD is a type of lung disease characterized by long-term breathing problems and poor airflow. The main symptoms include shortness of breath and cough with sputum production. COPD typically worsens over time, making even everyday activities difficult.
5. Trachea/bronchus/lung cancers (23): Lung cancer is characterized by uncontrolled cell growth in tissues of the lung. It can spread beyond the lung by the process of metastasis into nearby tissue or other parts of the body.

6. Diabetes mellitus (22): DM is the term describing a group of metabolic disorders leading to prolonged high blood sugar levels. Symptoms include frequent urination, increased thirst, and increased hunger. Serious long-term complications include cardiovascular disease, stroke, chronic kidney disease, foot ulcers, and damage to the eyes. Diabetes is due to either the pancreas not producing enough (or no) insulin (type 1 DM) or the cells of the body not responding properly to the insulin produced (type 2 DM). Insulin resistance is a condition often caused by excessive body weight and insufficient exercise.

7. Alzheimer's disease and other dementias (21): Dementia is a broad category of brain diseases that cause a long-term decrease in the ability to think and remember. 60–70% of dementia cases are caused by Alzheimer's disease (AD), a chronic neurodegenerative disease that usually starts slowly and worsens over time. Starting with short-term memory loss, subsequent AD symptoms can include problems with language, disorientation, mood swings, loss of motivation, and behavioral issues.

8. Diarrheal diseases (19): Diarrhea is defined as having at least three loose or liquid bowel movements each day. It can result in dehydration due to fluid loss. Signs of dehydration often begin with loss of the normal stretchiness of the skin and irritable behavior. This can progress to decreased urination, loss of skin color, a fast heart rate, and a decrease in responsiveness as it becomes more severe. The most common cause is an infection of the intestines due to either a virus, bacteria, or parasite—a condition also known as gastroenteritis. These infections are often acquired from contaminated food or water.

9. Tuberculosis (19): TB is an infectious disease usually caused by the bacterium *Mycobacterium tuberculosis* (MTB). Tuberculosis generally affects the lungs but can also affect other parts of the body. About 10% of latent TB infections progress to active disease which, if left untreated, kills about half of those infected. The classic symptoms of active TB are a chronic cough with blood-containing sputum, fever, night sweats, and weight loss. Tuberculosis is spread through the air when infected people cough, spit, speak, or sneeze.

10. Road traffic accidents (10): In 2013, 54 million people worldwide sustained injuries from traffic collisions, resulting in 1.4 million deaths.

Causes of death vary greatly between developed and less developed countries. For instance, high-income countries have decreasing traffic death rates, but higher cancer death rates.

In the USA,[58] the mortality rates are ranked as follows:

1. Heart disease
2. Cancer
3. Chronic lower respiratory diseases
4. Accidents (unintentional injuries)
5. Stroke (cerebrovascular diseases)
6. Alzheimer's disease
7. Diabetes
8. Influenza and pneumonia
9. Nephritis, nephrotic syndrome, and nephrosis (kidney problems)
10. Intentional self-harm (suicide)

In subsequent chapters we will describe various animal species with exceptional capabilities that (at least potentially) connect them to sapiens. Before moving there, a few more thoughts related to the human brain as compared to animals.

The Human Brain, The Big Differentiator

Because of our brain, human activities are forward-looking and purposeful. Human beings can use their own tools to engage in production activities and obtain their own necessary material information from the natural world, even making nature serve their needs and desires. Compared to animals, our human brain is the most important differentiator, enabling us to perform the following list of functions:

1. Carry out creative labor, manufacture and use production tools. Humans can use the tools they have created to change the physical forms of nature and the natural environment in which they live, in order to meet their own needs and suit their

own survival. They do not depend on changes in their own physiological structure. On the other hand, animals can only obtain ready-made food or other things provided by nature using their own physiological organs, such as teeth, tongue, and limbs, directly or in combination. Therefore, adapting the animals to the natural environment requires changes in their physiological structures.

2. Socialize and develop sophisticated group structures and networks. Humans are not the only social creatures in nature: ants and bees have a natural organization and dedication, work hard and are successfully dividing up labor. However, animal social structures are still simple compared to humans.

3. Rely on reason to control our own behavior, whereas animals only rely on instinct. We will further discuss this aspect below.

4. Communicate through spoken language. Humans gain experience in living knowledge through language and writing: They can communicate "lessons learned" so other will not repeat mistakes. Because social groups are safer and more convenient for language exchange, we form societies. It is because our brain has mastered language and written texts, we have evolved human civilization. Through language, people can quickly inherit the cultural achievements of their predecessors. We acquire new knowledge about the world by *learning*, and we retain the knowledge by storing it in memory and communicating it by word and written language. By doing so, sapiens has become the most wise and social animal. Because animal languages do not have as much clear and accurate information as human languages, the knowledge and experience they have accumulated cannot be effectively passed on to future generations, and knowledge cannot be accumulated. Animals must play the role of predator and prey, and wait for the food chain to shake and reshuffle the day.

5. Develop cultural inheritance methods based on language, writing, education, ideology, science, and various expressions of culture, including literature, music, painting, sculpture, and the performing arts. Human educational activities further increase the depth and breadth of cultural development. Learning is much more efficient than biological evolution.

Changes brought about by evolution are slow, requiring many generations. However, changes due to learning can be rapid, perhaps only requiring the life span of an individual. In the animal kingdom, any biological characteristics, life experiences and skills are accumulated and passed down, mainly through the genetic method of natural instinct. This method of accumulation and transmission is very limited and lots of experiences will be lost and cannot be inherited.

A concept closely related to the exceptional quality of the human mind is "consciousness." It has been defined in many different ways by philosophers since ancient times. Issues of concern in the philosophy of consciousness include whether the concept is fundamentally coherent; whether consciousness can ever be explained mechanistically; whether non-human consciousness exists, and if so, how it could be recognized. Our human mind has the unique ability to think about the future, to simulate various outcomes of contemplated actions and to pick the course of action deemed most favorable. Animals don't seem to be able to simulate the future; they live in the "now" and will behave in a certain situation as dictated by their "instinct."

Humans may also make decisions based on principles, on moral values that inform them about what's right and what's wrong. Another concept that has been discussed by philosophers throughout the ages is "free will." Are we guided by some superhuman power in our actions, or do we have the ability to decide for ourselves, to choose between alternative paths?

This is not to address those questions nor to try to attempt answers. What we are trying to point out is that only humans are struggling with such concepts that certainly go far beyond what animals can ever contemplate.

Let's end this sapiens chapter with an overview of some recent genetic findings that may help us understand what makes us human and why we have evolved our special cognitive functions.

US scientists recently compared the genomes of humans and other primates and found that this may be because the copy number of certain human genes is very different from other animals.[59] They compared more than 24,000 types of DNA. Using CGH (comparative genomic hybridization) technology, a total of more than 4000 gene copies representing species-specific changes were discovered. These genes with multiple copies often have very special effects. For instance, the gene called AQP7 and its expansion of the number of human germ line-specific gene copies can explain why humans have evolved *long-distance endurance*. The full name of AQP7 is aquaporin 7, and its role is to transmit water and glycerol across cell membranes. During intense human activity, AQP7 will catch human glycogen (i.e., energy) reserves. It also helps humans remove excess calories by sweating. The researchers also identified areas of the chromosome where some gene replication and gene loss are particularly active. It shows that these reproductions and losses are double-edged swords. It can provide a variety of alternative mutations for evolution, but it is also very easy to cause diseases. For example, the genes that can enlarge the gorilla diet will make humans susceptible to heart disease.[60] In addition, the researchers also found that the huge differences in the number of gene copies are potentially related to cognitive, reproductive, immune, and genetic disease susceptibility.

There are also studies that show that the main reason why human intelligence differs greatly from other animals is the "initial genetic mutation." As mentioned earlier, language is the key to distinguishing humans from other animals. So how did humans evolve languages? What is this gene that gives humans the gift of language? Scientists discovered that a mutation gene called FOXP2,[61] which is the evolution of this gene 200,000 years ago, has greatly improved human language ability. Scientists also discovered a gene called DUF1220[62] in the brain area. This gene is closely related to the brain's high cognitive ability. Scientists used computers to find the speed of genetic changes in humans, chimpanzees, and other vertebrates. They found that 49 discrete regions were related to evolution and found that HAR1,[63] the fastest-growing region, evolved 70 times faster than other regions. The region may have a close relationship with humans' advanced cognitive functions.

Brain research is one of the most active areas of science. Therefore, we can expect to learn much more about the inner workings of the brain and the genetic underpinnings of our special status as the most advanced of all living species.

Chapter 5

The Human Microbiome: How Our Health is Impacted by Microorganisms

Throughout human history, people have wondered about the relationship between human health, the food we digest and the microorganisms that reside on various human tissues and within our biofluids. However, our understanding of the relationship between microbes and health has long been limited. In 1908, Ilya Metchnikov shared the Nobel Prize with Paul Ehrlich for their work on immunity.[64] Honored for his discovery of phagocytosis (a cellular mechanism used to remove pathogens and cell debris), he is known as both the "father of innate immunity" and the "father of lactic acid bacteria." As the deputy director of the Pasteur Institute in Paris, Metchnikov got interested in "bifido-bacteria" (characterized by a Y-shaped "bifid" morphology) when they were first isolated in the intestinal flora of breast-fed infants. After observing that the longevity of Bulgarian peasants was associated with their consumption of fermented milk products, Metchnikov suggested that "oral administration of cultures of fermentative bacteria would implant the beneficial bacteria in the intestinal tract." He further developed the hypothesis that toxins produced by the human intestinal microflora cause human aging and disease.[65-67] He proposed that the human body's own poisoning and increased risk for disease are associated with human waste that is excessively deposited in the body. However, because of the lack

Our Animal Connection: What Sapiens Can Learn from Other Species (Second Edition)
Michael Hehenberger and Zhi Xia
Copyright © 2021 Jenny Stanford Publishing Pte. Ltd.
ISBN 978-981-4877-50-3 (Hardcover), 978-1-003-13072-7 (eBook)
www.jennystanford.com

of direct evidence, the relationship between intestinal microbes and human health could not yet be well explained. In 2001, another Nobel Laureate—the bacteriologist and geneticist Joshua Lederberg[68]— coined the term "human microbiome," thereby emphasizing the importance of nutrition for human health. His definition of the microbiome (or "microbiota") includes all members of the microbial community (bacteria, eukaryotic cells, viruses, etc.) that are housed in the human gut, the skin, the mouth, the nose, etc., as shown in Fig. 5.1.

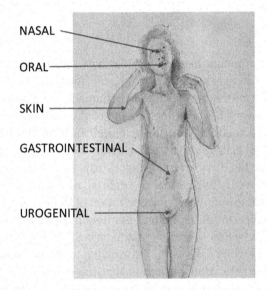

NASAL

ORAL

SKIN

GASTROINTESTINAL

UROGENITAL

Figure 5.1 The microbial community occupies 5 major sites in the human body (source: Wikipedia).

Collectively, microbiome members are surpassing the number of cells making up the human body, and there are 100–150 times more genes in the microbiome, as compared to the number of genes in the human genome.

There is mounting evidence that our microbial guests are playing an important role for our wellness, confirming that Metchnikov was up to something important when he more than 100 years ago suspected that the immune system is interacting with the microbiome. Without surface microbes, our bodies will not function properly. For example, intestinal probiotics can help digest food, suppress obesity, and fight invading microorganisms. Some microorganisms that colonize humans do coexist without harming humans; others have a mutualistic relationship with their human hosts. Certain microorganisms perform tasks that are known to be useful to the human host but the role of most of them is not well understood. Those that are expected to be present, and that under normal circumstances do not cause disease, are referred to as normal flora or normal microbiota. Depending on individual eating habits and other environmental factors, the intestinal microflora of individual persons is subject to considerable variation.

Along with the human body surface, the digestive tract is the conduit most exposed to the external environment. As shown in Fig. 5.2, it starts from the mouth and continues to the pharynx, esophagus, stomach, small intestine, large intestine, and ends in the anus (rectum).

The oral cavity serves as the main channel for feeding. The glands in the oral cavity secrete saliva. The food is lubricated by saliva through the esophagus. The saliva amylase in saliva decomposes some of the carbohydrates. The pharynx serves as a common passage for the respiratory tract and digestive tract and mainly completes the swallowing action. The esophagus is one of the muscular conduits, typically having a total length of 25 to 30 cm. The anatomy of the stomach is quite complicated and the total volume of the stomach is considerable, 1000–3000 ml. The mucous membrane contains a large number of glands that secrete gastric juice and become acidic. The small intestine is located in the abdominal cavity and its total length is 4 to 6 m. Its last part is equipped with a very strong digestive function. The total intestinal surface area is very large, up to 200~300 m^2, enough to accommodate a large variety of microorganisms. Note that the intestine is not only an important place for digestion and absorption of the human body, but also the largest *immune organ*. Being a good habitat for microorganisms, it can maintain normal immune defense function. The number of microorganisms in the intestine is larger than the number of human cells and their weight is estimated at up to 1.2 kg, about as heavy as the human liver. The human metabolic function does not cover all digestive needs and is therefore critically dependent on the microbiome.

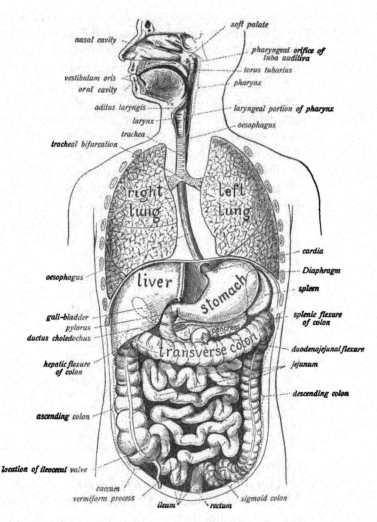

Figure 5.2 Human digestive system (Sobotta's human anatomy) (source: Wikipedia).

A recent study[69] found that there are at least 1000 to 1200 bacterial species in the human intestine. Each host contains approximately 160 dominant bacteria. The human intestine contains not only harmful bacteria but also various beneficial bacteria. The human intestine normally contains microorganisms such as bifidobacteria, lactobacillus, etc. Most of them are strict anaerobic bacteria, a small number of them are facultative anaerobic bacteria (i.e., capable of making ATP by aerobic respiration if oxygen is present, but also able to switch to fermentation or anaerobic respiration if oxygen is absent). The gut bacteria of most healthy adults are relatively stable, while the elderly often suffer a reduction of the number of bifidobacteria. Once a person suffers from a certain disease, the bacterial structure will undergo changes. Those changes can eventually result in a condition called dysbacteriosis, a term that describes microbial imbalance or a seriously impaired microbiota. For example, a part of the human microbiota, such as the skin flora, gut flora, or vaginal flora, can become deranged, with normally dominating species underrepresented and normally outcompeted species increasing to fill the void. The healthy person's gastrointestinal tract accumulates a wide variety of microorganisms. These microorganisms are called intestinal flora. Under normal circumstances, a large number of bacteria in the intestinal tract exist in a certain proportion. Each group of bacteria has mutual restraint, interdependence, to other bacteria. There is a specific ecological balance between the groups. Although the environment and organisms constantly change the flora, the bacteria and the organism will always maintain a corresponding dynamic balance. Once the host's internal and external environment changes, especially by the frequent use of broad-spectrum antibiotics, the susceptible flora in the gut will be inhibited, and other uninhibited bacteria may multiply, thereby causing dysbacteriosis, i.e., making the patient's disease worse.

Among the microorganisms in the human intestine, more than 98% are bacteria. They can be roughly divided into three categories.

(1) **Probiotics**: also known as probiotics, mainly bifidobacteria, lactobacillus, etc., are indispensable for human health and represent the main constituents of the intestinal flora. They can synthesize various microorganisms and participate in digestion and absorption of food, promote intestinal peristalsis, inhibit the production of some harmful toxins and gases, help the body to improve immunity and maintain the balance of intestinal flora.[70]

(2) **Harmful bacteria**: Generally present in the intestines and harmless to the human body, those bacteria can grow out of control and reproduce in large numbers. Under certain

conditions, such imbalances will lead to damage to the patient and cause a variety of unpleasant conditions: defecation may no longer be smooth; a large amount of manure may accumulate in the intestines, producing harmful substances and reabsorbing substances that are harmful to the body.

(3) **Neutral bacteria**: Bacteria with double functions, such as *E. coli*, are generally beneficial to the body. Once their numbers are out of control, they become harmful bacteria and can be transferred from the intestinal tract to other parts of body where they may cause a variety of diseases.

The ecological balance between the human body and the microorganisms is regulated by host factors and environmental factors. Beneficial bacteria keep their own stability by producing bacteriocins, antibiotics, and other metabolites. Most of the ecological disorders are caused by factors such as chronic diseases, cancer, surgery, and radiation.[71,72] The age of the organism also affects the balance of intestinal flora to a large extent. As the age of the host increases, the organs and functions of the organism gradually deteriorate. The beneficial microbial flora in the gut also gradually decreases, and the harmful microbial flora gradually increases. Healthy young adults, on the other hand, are still in a balanced state of the intestinal microflora, and the beneficial microorganisms are clearly dominant. Healthy beneficial bacteria are therefore playing an important role in maintaining the microbial balance of the intestine. They play an important role in people's life and health. It has been found that people who regularly eat beneficial bacteria have a longer lifespan.[73] In a word, maintaining the number of beneficial bacteria in the intestinal tract is beneficial to maintaining the ecological balance of the flora in the gut, and is also conducive to the maintenance of a person's health.

The bacteria in the intestinal tract of healthy individuals are combined at a certain proportion, and each bacterium is mutually restricted and interdependent, maintaining a certain amount and proportion of ecological balance. Once the internal and external environments change, the ecological balance is destroyed. This happens due to frailty, critical illness, immunosuppressive therapy and (radio)-chemotherapy, and applies particularly to patients with long-term extensive use of antibiotics.[74] Intestinal beneficial bacteria are inhibited, pathogenic bacteria grow, causing dysbacteriosis and often showing symptoms such as abdominal pain and diarrhea. Consequently, the human large gut microbiota has evolved into a vital "organ" that is inextricably linked to the human body through long-term co-evolution with the host. The gut microbiota "organ" performs a variety of functions, including substance metabolism, biological barriers, immune regulation, and host defenses. Gut microbes not only help the body absorb nutrients from food, but also synthesize amino acids, organic acids, vitamins, antibiotics, etc. They can metabolize the produced toxins to reduce the toxicity to the human body. Different eating habits and lifestyles have a great influence on the type of human intestinal microflora. For example, a high-fat diet can lead to reduction or even disappearance of beneficial bifidobacteria. There is a mutually beneficial relationship between the intestinal microbes and the human body. Our gut microbiota plays an important role in maintaining human health.[75]

In addition to substance synthesis and metabolic functions, the relationship between the complex microbial ecosystem of the intestine and the body's immune system is also extremely close. Intestinal microbes can not only serve as natural barriers to maintain the integrity of the intestinal epithelium, prevent the invasion of pathogenic microorganisms, but also regulate the intestinal mucosal secretion of antibodies on the intestinal immune system. They are affecting both our natural immune system and acquired immunity. The gut microbiota is considered to be the largest "immune organ" in the human body. The immune balance maintained by the gut microbiota plays an important role in the prevention of autoimmune diseases. When certain factors cause changes in the intestinal microbiota, it will further affect other immune systems of humans. Once this immune balance is broken, it will easily lead to the emergence of various diseases. More and more experimental evidence shows that the intestinal microbes not only affect the function of the human intestine itself, but also affect human health from different angles and levels by regulating the human immune system.

Under natural conditions, babies are born through the birth canal of pregnant women. From the initial "home" to the "outside" process, a large number of microorganisms enter the digestive tract through the baby's mouth and colonize their digestive tract, and then pass through the skin of their mother. During the process of contact, breast-feeding, the baby's initial intestinal microbiota "inherits from her mother." With the growth of the baby, the intestinal microbiota gradually continues to build, and it enters a relatively stable state during the adult period. Intestinal microorganisms such as bifidobacterium and lactobacillus can synthesize various vitamins necessary for human growth and development,[76] and can also use the degraded protein residues to synthesize necessary amino acids to participate in the metabolism of carbohydrates and proteins. They can also promote the absorption of mineral elements such as iron, magnesium, and zinc.

Diabetes and Intestinal Microbes

Type 1 diabetes is an auto-immune disease and the gut microbiome is highly suspected to play a role in triggering the destruction of insulin-producing beta cells in the pancreas. T1D patients frequently report that the onset of their disease was associated with an infectious disease (such as laryngitis) that had been treated with antibiotics, thus creating microbiota imbalance. A significant increase in the number of T1D patients has been reported in countries with high levels of hygiene, comprehensive immunization programs and frequent use of antibiotics—all factors which impact the microbiome.

As to Type 2 Diabetes, BGI collected 345 samples of T2D patients in China and sequenced and analyzed intestinal microbes. Through in-depth analysis of type 2 diabetes-related microorganisms, the researchers found that there is an antagonistic relationship between beneficial bacteria and harmful bacteria, especially the association between Clostridium and diabetes, which suggests that patients that are missing specialized butyrate-producing bacteria are susceptible to diabetes. The concept of the metagenomic linkage group (MLG) was proposed. In addition, intestinal microbes can be used to assess and monitor the risk of type 2 diabetes and other diseases. However, at a later stage of the disease patients do no longer exhibit a particular intestinal type.[77-79]

Intestinal Microorganisms and Immune System

The relationship between intestinal microbes and the immune system can also be shown by lab experiments with mice. While mice growing up in a sterile environment have immunodeficiencies, by supplementing the gut bacteria, the immune deficiency can quickly recover.

Taking rheumatoid arthritis as an example, the analysis of fecal microorganisms in patients with rheumatoid arthritis showed that rheumatoid arthritis was closely related to a species of bacteria known as *Prevotella copri*. The abundance of those bacteria is related to the reduction or disappearance of some beneficial bacteria.

An article published in *Nature* reported that intestinal commensal[a] bacteria can regulate the differentiation of multiple T cells and thus change the immune system of the intestinal mucosa. *F. prazilus* is located in the mucosal layer of the intestine and produces butyrate by fermentation. This short-chain fatty acid stimulates and modulates T cells to prevent the development of intestinal inflammation. All Clostridium bacteria have similar mechanism. Another article in *Science* pointed out that under normal circumstances, dendritic cells do not respond to T-cell inflammation in the intestinal mucosa, so they play an important role in maintaining intestinal immune tolerance.[80] However, when the environment changes, dendritic cells can activate T cells, and β-chain proteins on T cells play an important role in regulating dendritic cells. When the β-chain protein is cleared, the activity and resistance of T cells are regulated. The effect of inflammatory cytokines was significantly reduced, while pro-inflammatory helper T cells 1 and 17 and their cytokines increased. Mice lacking beta-catenin in dendritic cells exhibit increased sensitivity to enteritis.

In fact, the gut microbiota and the immune systems are two sets of interacting systems that have undergone long-term evolution in nature. It can be said that the interaction between the human body and the gut bacteria has largely ensured that the human body has a complete set of immune system. *Together*, they can resist various pathogens from the outside world.

---

[a]Commensal bacteria obtain benefit from another organism without affecting it.

The Role of the Appendix

The human appendix, a narrow pouch that is part of the digestive system, has a bad reputation for its tendency to become inflamed (appendicitis), often resulting in surgical removal. The worm-shaped structure found near the junction of the small and large intestines is widely viewed as an organ with little known function. However, recent research[81] suggests that the appendix may serve an important purpose: it serves as a reservoir for beneficial gut bacteria and is actually an immune organ that helps the lymphatic system detect and eliminate pathogens. The appendix produces a special type of immune cell, the intrinsic lymphocyte. It is a kind of immune effector cells that was formed during the long-term evolution of organisms.[82] It can respond quickly to invading pathogens and produce nonspecific anti-infective immunity, which plays an important role in protecting people with low immunity against bacterial infection. It can also participate in the elimination of damaged, aging or distorted cells in the body, and participate in adaptive immune responses. Intrinsic lymphocytes[83] also help the appendix to store beneficial bacteria to maintain the balance of the intestinal flora in the human body.[84]

Another recent (2016) study by Duke University[85] confirmed that a large number of bacteria that are beneficial to the human gut are stored in the appendix. When a person suffers from problems in the intestinal tract due to illness or other causes, the appendix will contribute these bacteria to help the intestine reshape the healthy flora system and maintain the balance of intestinal microbes. In other words, the appendix acts like a safe "backup disk" for the intestinal flora, a reservoir of beneficial bacteria. It can reduce the damage caused by pathogenic bacteria in the human body.

When humans suffer from stomach infection or even food poisoning, the balance of beneficial bacteria in the human body can play a key role in resisting the pathogenic virus.

However, removal of the appendix may have at least one benefit: according to a new study[86] based on a large population of patients, it was that people who had had appendectomy in the early years had a 19.3% reduction in the risk of Parkinson's disease, and the onset time was an average of 3.6 years later. Further studies have shown that this is due to the role of α-synuclein in the appendix. The pathological protein α-synuclein is suspected to trigger Parkinson's disease.

Intestinal Microorganisms and Antibiotic Use

Antibiotics can be used to fight bacterial infections in the body. It can be seen that intestinal microbes are also affected by antibiotics. However, the effect is still unknown. We know that early neonatal antibiotic treatment can have a systemic effect on infants and cause problems in the construction of neonatal immune systems. At the same time, different antibiotics have different effects on different intestinal microbiota. It is clear that antibiotic treatment will break the original intestinal microbial balance and reduce intestinal microbial diversity. Repeated use of antibiotics can also produce resistance, making it difficult to treat endogenous bacterial infections.

Intestinal Microbiology and Neurodevelopment

In recent years, there has been an increased interest in the effects of intestinal microbes on neurodevelopment. An article in *Cell* showed that scientists have discovered intestinal microbiological disturbances in mice with autism, and they also found that they can be supplemented with bacteroides fragilis for treatment.[87]

Another interesting experiment comes from the study of "anxiety."[88] The researchers transplanted the intestinal microbes of "anxious" mice, resulting in "not anxious" mice exhibiting "anxiety" behavior.

The effects of gut microbes on neural development may have already been completed during early neurodevelopmental processes,[89] because their effects on *adult* mice are very limited. This highlights the critical period of neural development in which gut microbiota plays a role. It appears that any defects occurring during early development will be lasting.

Intestinal Microorganisms and Obesity

Whether or not obesity is a disease is controversial, but the diseases caused by obesity are indisputable facts. Although research studies have revealed differences in the types and numbers of intestinal flora among obese and non-obese people, there has been no direct evidence of whether obesity is caused by specific bacteria.[90] German scientists have found the culprit leading to obesity in mice: a bacteria called *Clostridium ramosum*. They fed mice with high-fat and low-fat diets for 4 consecutive weeks. As a result, the proportion of clostridium perfringens in the caeca of mice fed with high-fat diets was approximately 4 times higher than that of low-fat diets. They speculated that the cause of obesity is that it can increase the absorption of food nutrition. However, whether research results obtained with mice are applicable to humans will require further investigation.

Intestinal Microorganisms and Hypertension

Studies have found that short-chain fatty acids (SCFAs) produced by intestinal microbes are associated with blood pressure regulation.[91] Through the study of two major SCFAs-knockout mice (a term to be explained in the "mouse chapter" below), scientists found that antibiotic treatment of knockout mice resulted in decreased gut microbiota and increased blood pressure in mice. Propionate is a type of SCFA, and when given propionate, the blood pressure of mice undergoes a rapid and dose-dependent decline, and knockout propionate receptor mice are particularly sensitive to this effect. In addition to short-chain fatty acids, some substances produced by intestinal probiotics such as gamma-aminobutyric acid (GABA) also have the effect of lowering blood pressure.

Intestinal Microbes and Chronic Inflammatory Bowel Disease

Under normal conditions, the intestine does not produce fructose. A study published by *Nature* found that the intestinal tract of mice with chronic inflammatory bowel disease actively produces fructose, providing nutrition for the growth of probiotics to fight against the growth of harmful bacteria.[92] The mechanism by which the gut actively nourishes probiotics in mice derives from a function that encodes a sucrose synthase gene. When mice are lacking this gene, fructose synthesis capacity is lost, growth of probiotics is impeded, and pathogenic bacteria slip in, eventually leading to chronic inflammatory bowel disease.

Intestinal Microorganisms and Tumors

A high-fat diet changes the intestinal microbiota and stimulates the growth of intestinal tumors. Here we take liver cancer as an example to illustrate the relationship between intestinal microbes and tumor formation. Japanese scientists have found that the formation of liver cancer is closely related to obesity and intestinal microbes. They fed obese and normal mice/rats with chemicals that cause liver cancer. As a result, all obese mice developed liver cancer, and only 5% of normal rats developed liver cancer. After the use of antibiotics to remove intestinal microbes, the incidence of hepatocellular carcinoma in obese rats is greatly reduced.[93] Further studies have found that some enteric bacteria belonging to the phylum Streptococcus produce lipopolysaccharides, some of which are transformed by bacteria into toxic deoxycholic acid (DCA). The concentration of DCA in the blood of obese mice is high, and these lipopolysaccharides and DCA accumulate after reaching the liver, causing inflammation and DNA damage, eventually leading to liver cancer. Compared with normal mice, obese mice produce more DCA intestinal microorganisms, which makes it more likely that obese mice will develop liver cancer. The increase in DCA concentrations is therefore coupled to the incidence of liver cancer.

Intestinal Microbes and Neurobehavioral Diseases

Recent studies have shown that gut microbes can regulate a host of behaviors of the host through the gut-brain axis.[94] Taking autism as an example, children with autism often have severe gastrointestinal disorders in their clinical manifestations. A large-scale survey conducted by the US Centers for Disease Control and Prevention shows that children with autism have a higher proportion of gastrointestinal diseases. Scientists at Caltech and other collaborating universities[b] fed *Bacteroides fragilis* to mice with autism, improving their intestinal permeability and making their intestinal microbial population closer to normal mice. They found that those mice became more flexible and had their anxiety and other symptoms reduced. Metabolomic analysis of mouse serum revealed that gut bacteria affect the behavior of mice by affecting the levels of some metabolites. Injecting one of these metabolites into normal mice can lead to abnormal behavior.

Intestinal Microbiology and Biological Clock

Israeli and German scientists[c] conducted an interesting experiment: First of all, they analyzed the fecal microbiota of mice that maintained normal circadian rhythms and found that up to 60% of the absolute and relative proportions of intestinal bacteria caused fluctuations in circadian rhythms. They further analyzed the feces of normal life rhythms and people who frequently traveled to the United States and Israel for long distances. They found that the characteristics of intestinal flora in these individuals were very similar to those of mouse experiments before, during, and after the jet lag. Changes in the characteristics of obese and diabetic patients have occurred in the intestinal flora of people with jet lag. More interestingly, the transplanted feces of the subjects during the jet lag process were transplanted into healthy mice. These mice gained weight, and their blood glucose and body fat were higher than those of the control group. This also shows that the habit of staying up late may have a negative impact on health and that regular sleeping habits are good for health.

From the latest research on the development of intestinal microbes and diseases, it is not difficult to see that intestinal microbes play an important role in maintaining human health. Many diseases are directly or indirectly related to the imbalance of intestinal microbes. In addition to the direct relationship between the onset of the disease and the intestinal microflora, constipation, bad breath,

[b]https://www.sciencedirect.com/science/article/pii/S0092867413014736
[c]https://www.ncbi.nlm.nih.gov/pmc/articles/PMC4598175/

mouth ulcers, endocrine disorders, and other problems that we often encounter in our daily lives are also related to the imbalance of intestinal microbes. Because of infection with a certain pathogenic microorganism, or after the use of antibiotics, human patients often experience symptoms such as constipation, bitter taste, bad smell in the mouth, and poor appetite. The reason is that antibiotics not only kill pathogenic microorganisms, but also eliminate some beneficial microorganisms, change the composition of the intestinal microbes, causing people to feel bad.

Intestinal Microbes and Human Mental Health

The nerve cells in the intestine are very similar to the neurons in the brain regarding cell types, neurotransmitters, and receptors, and are equal to the number of nerve cells in the brain, exceeding their number in the spinal cord. Therefore, the intestine is also called the human "second brain" or "intestine brain." The intestine brain and the brain are in two-way communication and they are connected by a brain–gut axis. The intestinal brain can affect the central nervous system, and then affect the human brain's cognitive function and behavior, in which intestinal microbes may play a very important role. Autism may also be affected by the brain–gut axis. The intestine is the largest digestive organ in the human body and is responsible for the digestion and absorption of nutrients. Small molecules can be absorbed directly by the intestine. Certain macromolecules need to be absorbed by the body through specific channels or receptors.

Because disorders of the intestinal flora often cause diseases of the central nervous system, they are involved in irritable bowel syndrome, inflammatory bowel disease, and hepatic encephalopathy.[95] They contribute to a large number of gastrointestinal symptoms while causing inflammatory diseases. They can trigger mental disorders in the patient such as anxiety, depression, nervousness and other emotions. Clinical studies have shown that dysbacteriosis can often lead to a variety of neurological diseases such as autism and Alzheimer's disease. It has been found that the gut microbiota composition in patients with gut-brain disease and central nervous system disorders has changed significantly compared with healthy controls, and that dysbacteriosis in the gut flora will lead to animal-like intestinal–brain diseases and central nervous system disorders, suggesting that dysbacteriosis may be the cause of bowel–brain disease and central nervous system diseases.[96,97]

Since the relationship between gut microbes and health is so important, how can we maintain the health and prevent and treat some diseases by regulating intestinal microbes? Some well-known probiotics can regulate intestinal microorganisms. Yoghurt is fermented by probiotics. It not only helps to improve body weight, but also relieves irritable bowel syndrome, depression, and hypertension. A recent large-scale follow-up survey showed that eating a certain amount of yogurt every day can greatly reduce the prevalence of diabetes, although its mechanism of action is still not clear. It is believed that the probiotics in yogurt can help increase insulin sensitivity and reduce inflammation.

In addition to probiotics, human feces can also be used to treat diseases. We have recently seen news that companies in the United States have established fecal banks. They actually collect intestinal microbes from healthy people to treat intestinal infectious diseases. For example, due to the large number of deaths of other bacteria caused by the administration of antibiotics, *Clostridium difficile* is over-proliferated in the intestine, causing abdominal pain, diarrhea and other symptoms. The feces of healthy persons are collected and sent to the patient's intestine through various methods after treatment. The normal intestinal flora contained in these fecal extracts can inhibit the over-breeding of *C. difficile*. The results of clinical trials in the United States and Canada have proved that the effective rate of this method may exceed 90%.

The potential benefits of gut bacteria analysis are not only restricted to humans, they may also help with conservation efforts of endangered species: after analyzing 200 different koalas at 20 sites, scientists found[98] that some had a very restricted eucalyptus diet because of their inability to digest an available eucalyptus strain. After fecal transfers of gut bacteria from other koalas, the vulnerable koalas were able—after only 2 weeks—to eat the previously undigestible eucalyptus strain. This finding could save species with limited food supply by extending their available diets.

In addition to using our own intestinal microbial resources, the goal of improving the function of the gut microorganisms by genetically engineering microorganisms has also been met with some encouraging progress. U.S. researchers expressed a gene that suppresses appetite in a kind of probiotic bacteria. They then let mice that were fed a high-fat diet, drink water containing this probiotic. After eight weeks, the food intake of mice was significantly reduced, along with reduced fat and insulin resistance. The degree of fatty liver was significantly lower than those of the mice that received the control strain, and the same effect could be maintained four weeks after the cessation of the bacterium. It is believed that the use of similar methods to prevent and treat the disease will no longer be a distant matter. In addition, the researchers envision the use of bacteriophages against a particular bacteria as a means of treating disease by regulating the proportion of harmful bacteria in the gut. Recently, American scientists have discovered a bacteriophage whose host is a Bacteroid related to diabetes and some intestinal diseases. Because phage specificity and ease of entry into the intestine play a role, it is relatively safe and effective.[99,100]

Although our understanding of the relationship between intestinal microbes and human health has evolved, many problems still need to be solved. For example, a large number of gut microbes cannot yet be cultivated alone. However, it is believed that in the near future, the gut microbiological code that affects the health of microbes will be cracked one by one. Several international research collaborations have been established, among them the European Metagenomics of the Human Intestinal Tract (MetaHIT[101]) project, the NIH-funded Human Microbiome Project[102] (HMP), and the Earth Microbiome Project[103] (EMP). Realizing a healthy body by regulating intestinal microbes will no longer be a dream, as interdisciplinary research into zoology, microbiology, bacteriology, genetics, and medicine will generate breakthroughs that promise to have a revolutionary impact on human health.

Chapter 6

Animals with Connection to Human Knowledge, Health, and Performance*

The animals covered in this book are grouped into three categories:

A. Model organisms studied to increase human knowledge
B. Animals that offer ways to improve human health
C. Animals studied because of their potential to improve human performance

6A. Human Knowledge

Human knowledge in fields such as genetics, molecular biology, cellular biology, physiology and other life sciences disciplines, is gained and enhanced by studying "model organisms." Since the completion of the Human Genome Project (HGP) and the realization that there are many similarities between human and animal genes, research into "model organism biology" has received even more attention. Model organisms identified by HGP can be grouped roughly into seven categories: *bacteria, yeast, worm, fly, weed, fish,* and *mouse*. Note that the list includes a plant ("weed"), forcing us to briefly jump outside the scope of this book.

The knowledge accumulated by the global scientific community about model organisms is typically shared in Model Organism

*Refer to Chapter 7 for more information.

Our Animal Connection: What Sapiens Can Learn from Other Species (Second Edition)
Michael Hehenberger and Zhi Xia
Copyright © 2021 Jenny Stanford Publishing Pte. Ltd.
ISBN 978-981-4877-50-3 (Hardcover), 978-1-003-13072-7 (eBook)
www.jennystanford.com

Databases (MODs). They are dedicated to the provision of in-depth biological data for intensively studied model organisms. MODs allow researchers to easily find background information on large sets of genes, plan experiments efficiently, combine their data with existing knowledge, and construct novel hypotheses. They allow users to analyze results and interpret datasets. Where possible, MODs share common approaches (such as gene ontology[104]) to collect and represent biological information. "Gene ontology" is used to describe functions, processes and cellular locations of specific genes. It is a major global initiative to unify the representation of gene and gene product attributes across all species. More specifically, the project aims to (i) maintain and develop its controlled vocabulary; (ii) annotate genes and gene products, and assimilate and disseminate annotation data; (iii) provide tools for easy access to all aspects of the data provided by the project. The scientific community is further sharing software for the curation, visualization, and querying between different MODs. Model Organism Databases are also helpful for projects focused on less well studied species.

HGP specific Model Organisms

Bacteria: E. coli

As already discussed previously (Chapter 5), *Escherichia coli*,[a] also known as *E. coli*, is a gram-negative,[105] facultative aerobic, rod-shaped member of the coliforms group of bacteria. Gram-negative bacteria are defined as bacteria that do not retain the crystal violet stain used in the gram-staining method[b] of bacterial differentiation. *E. coli*, shown in Fig. 6A.1, is commonly found in the lower intestine of warm-blooded organisms (endotherms).

E. coli is one of the commonly used indicators of food contamination, and it is the culprit in causing diarrhea. There are good bacteria in the family that help to produce some vitamins in the intestines, as well as bad ones that can make people sick. The coliform group and the total number of colonies are the indicators for judging

---

[a]Named after Austrian pediatrician Theodor Escherich, who discovered *E. coli* and studied its properties.
[b]Named after Danish bacteriologist Hans Christian Gram, who developed the technique.

the degree of food contamination. We often hear that certain foods are not qualified because of the high coliforms, indicating that this food is very likely to contain a large number of pathogenic bacteria. The risk of food poisoning after eating will then be high.

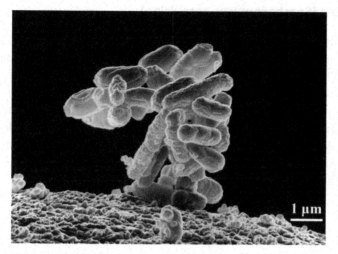

Figure 6A.1   Rod-shaped bacterium *Escherichia coli* (source: Wikipedia).

The genera Escherichia and Salmonella diverged around 102 million years ago, coinciding with the divergence of their hosts: Escherichia being found in mammals and Salmonella in birds and reptiles. Salmonella species are intracellular pathogens: certain types cause illness. They usually invade only the gastrointestinal tract and cause salmonellosis; symptoms resolve without antibiotics. Typhoidal serotypes can only be transferred from human-to-human, and can cause food-borne infection including typhoid fever. Typhoid fever occurs when Salmonella invades the bloodstream. A septic form of Salmonella can spread throughout the body, invades organs, and secretes endotoxins, leading to life-threatening septic shock and requiring intensive care.

E. coli is a model organism of prokaryotes. It has been a classic experimental material for molecular genetics and molecular biology, and later became the main engineering cell of genetic engineering. The genes and expression regulation of *E. coli* are clear, and the genetic map is complete and exquisite. In September 1997, the genome-wide sequence map of *E. coli* was completed. The genome has a size of about 5 Mb and a total of 4288 coding genes with a gene density of about 857 genes/Mb. The correlation between the structure and function of most genes has been experimentally proven.

Yeast (Saccharomyces cerevisiae)

Yeasts are eukaryotic, single-celled microorganisms classified as members of the fungus "kingdom." During fermentation, as shown in Fig. 6A.2, the yeast species *Saccharomyces cerevisiae* converts carbohydrates (such as sugar) to carbon dioxide and alcohols, and both end products have been used by sapiens for thousands of years: the carbon dioxide has been used in baking, and the alcohol in alcoholic beverages. Yeast is also a centrally important model organism in biology and is one of the most thoroughly researched eukaryotic microorganisms.

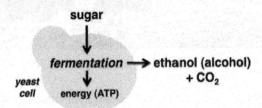

Figure 6A.2 Yeast enables fermentation, resulting in alcohol and CO_2 (source: Wikipedia).

In brewing beer, yeast in the anaerobic environment will "eat" the sugar in the grain (i.e., starch that is hydrolyzed in steamed or soaked wort), it will metabolize alcohol and carbon dioxide, as well as undergo a series of complex biochemical reactions that add a special flavor.

Making wine is even more complex: when the grapes are ripe, they will naturally ferment, because there are *Saccharomyces cerevisiae* on the peel of the ripe grapes. Few microbes are more important to humans than yeast. In the wine containers from 5000 years ago that were discovered by archaeologists, the DNA of yeast was also found. In 2012, scholars from the University of Florence, Italy, found that the intestines of the wasps are the sanctuary of *Saccharomyces cerevisiae*.[106] Yeasts on the outside of grapes and wasps have evolved a stable symbiotic relationship that allows the wasps to digest food and carry yeast in their intestines. At the time of pollination, these yeasts remain on the vines and can be used by humans in winemaking.

Saccharomyces cerevisiae (see Fig. 6A.3) is a model organism of single-cell eukaryotes.

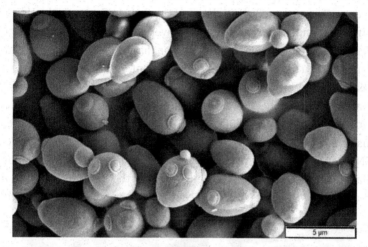

Figure 6A.3   Yeast (*Saccharomyces cerevisiae*) (source: Wikipedia).

Yeast has been a classic experimental material in genetics and molecular biology, and the main engineering cell of genetic engineering, especially in the biological industry. Being a eukaryotic organism, more than 30% of the coding genes of yeast have high homology with mammals and humans, so the yeast genome has greater significance for the Human Genome Project than *E. coli*, especially for the recognition of genes and other functional factors. Yeast is used in the elucidation of metabolic pathways and signaling pathways. The yeast genome sequence was published in October 1996. It consists of 16 chromosomes with a size of about 12 Mb, encoding about 6000 genes, and a gene density of about 500 genes/ Mb.

Worm (*Nematode,* Caenorhabditis elegans)

C. elegans is a free-living (not parasitic), transparent *nematode* (roundworm), about 1 mm in length, that lives in temperate soil environments.

Nematodes, also known as roundworms in animal taxonomy, are named after their shape, shown in Fig. 6A.4. *C. elegans* is a model organism of multicellular eukaryotes (invertebrates). The use of *C. elegans* as a model organism was proposed in the 1970s by Sydney Brenner at the Medical Research Council (MRC) Unit in Cambridge, UK. Sydney Brenner and John Sulston of the Sanger Center in the United Kingdom, and Robert Horvitz, MIT, were awarded the 2002 Nobel Prize[107] in medicine or physiology "for their discoveries concerning genetic regulation of organ development and programmed cell death," based on years of research with *C. elegans*. Andrew Fire of Stanford University and Craig Mello, University of Massachusetts, Worcester, received the 2006 Nobel Prize in Medicine[108] for "for their discovery of RNA interference: gene silencing by double-stranded RNA," again based on nematode roundworm research. In addition, Martin Chalfie, one of the three winners of the 2008 Nobel Prize in chemistry,[109] used green fluorescent protein in nematode research.

Figure 6A.4 *C. elegans* as photographed by William Wergin and Richard Sayre (source: Wikipedia).

The nematode is easy to culture and can be grown on agar medium and fed on *E. coli*. It can be stored in low temperature liquid nitrogen for several years. Adult nematodes are about 1.5 mm long and hermaphroditic. The adult organism is composed of a well-defined number of (959) somatic cells. Each cell has a clear morphological, developmental, and genetic background. The growth cycle is short, and the fertilized embryo can hatch into a free-living larva within 12 hours. The larva can mature in 40 hours. Adults can produce hundreds of offspring in about 4 days. Mutants were easily obtained in various ways, and the phenotypic characteristics were obvious. The genomic study of nematodes began in 1990, and genome sequencing and analysis were completed in 1998 and published on December 24 of that year. The *C. elegans* genome is about 100 Mb in size and consists of 6 chromosomes with about 20,000 coding genes. 60% of the genes are highly homologous to other eukaryotes.

At the same time, *C. elegans* also has a very unique gender system. Its two genders are male and hermaphroditic. The hermaphroditic nematode itself contains both testis and ovaries, so it can achieve autologous reproduction without males. Males only have a reproductive system, so they cannot achieve autologous reproduction and can only be propagated by hermaphrodites. This is of great significance for our experimental research using nematodes. In general, by cultivating hermaphroditic nematodes and using their own reproduction, we can ensure that the nematode genotype is stably inherited and will not change; on the other hand, presence of a male helps to introduce other genotypes and facilitate hybridization.

Fruit Fly (Drosophila melanogaster)

Drosophila (Fig. 6A.5) is a model organism of invertebrates (Insecta). It will be featured in more detail in a special chapter below. It has four pairs of easily identifiable chromosomes. There are many well-defined phenotypes, and the trait variation is obvious. It became the most widely used classic model animal in the early 20th century.

Drosophila was also the second multicellular organism to complete genome sequencing. The Drosophila genome has a total length of 140–180 Mb and a coding gene number of 13,792. DNA sequence of 116 Mb in the euchromatin region of the Drosophila genome was published in March 2000. Interestingly, 177 of the 289 genes that determine human genetic disease can be found as homologous genes in the Drosophila genome. Comparing the genome sequences of Drosophila with yeast and nematodes, the number of proteins in Drosophila and *C. elegans* was similar, but twice as high as that of yeast.

Figure 6A.5   *Drosophila melanogaster* (source: Wikipedia).

Thale Cress (Arabidopsis thaliana)

The thale cress, mouse-ear cress or Arabidopsis (see Fig. 6A.6), is a small flowering plant native to Eurasia and Africa. It is considered a weed, found by roadsides and in disturbed land. A winter annual with a relatively short life cycle, *A. thaliana* is a popular model organism, used as a tool for understanding the molecular biology of many plant traits, including flower development and light sensing. Since the 1980s, Arabidopsis has become an ideal experimental material for plant genetics, physiology, biochemistry, and development. HGP chose Arabidopsis as the only model plant, first because of its small genome, and secondly to compare the similarities and differences between plant and animal genomes. As a small dicot (the name *dicot* refers to one of the typical characteristics of this group of plants, namely that the seed has two embryonic leaves or *cotyledons*), Arabidopsis was the first genome to be sequenced and analyzed. The Arabidopsis genome was published in 2000. For a complex multicellular eukaryote, *A. thaliana* has a relatively small genome. The Arabidopsis genome has five chromosomes with a size of approximately 115.4 Mb and approximately 25,498 coding genes.

A. ROCKENTRAV, TURRITIS GLABRA L.
B. BACKTRAV, ARABIDOPSIS THALIANA (L.) SCHUR.

Figure 6A.6 Thale cress (*Arabidopsis thaliana*) (source: Wikipedia).

Pufferfish (Tetraodontidae)

Pufferfish (Fig. 6A.7) is a model organism of vertebrate (fish). Its inclusion as a HGP model organism should also be attributed to Sydney Brenner, who selected the pufferfish because of its small genome. The genome composition is highly similar to humans, but it lacks many introns and intergenic repeats.

It has great reference value in identifying coding genes and other functional factors and understanding the structure and evolution of vertebrate genomes. The puffer fish genome was published in August 2002. It is only 392 Mb in size and has 21 chromosomes, encoding about 31,059 genes.

Figure 6A.7   Pufferfish (source: Wikipedia).

The majority of pufferfish species are toxic and some are among the most poisonous vertebrates in the world. The internal organs, such as liver, and sometimes their skin, contain tetrodotoxin and are highly toxic to most animals when eaten; nevertheless, the meat of some species is considered a delicacy in Japan (Japanese: fugu), when prepared by specially trained chefs who know which part is safe to eat and in what quantity.

Mouse (Mus musculus)

Mice are model organisms of vertebrate (mammal). Mice (Fig. 6A.8) are the most commonly used and most important medical laboratory animals and will be featured below in a special chapter of this book. HGP selected mice as model organisms because of the size and location of the genome, the structure and location of chromosomes or segments, the density and distribution of coding genes, and other functional factors. The composition of repeat sequences is highly similar to humans, with more than 90% of mouse genes finding corresponding homologous genes in the human genome, which is of great significance for the assembly and annotation of the human genome.

In December 2002, the sequence of the C57BL/6J strain mouse genome was "sketched" with a genome size of about 2.5 Gigabases (Gb), an estimated number of genes of about 30,000.

Figure 6A.8   Mouse (*Mus musculus*) (source: Wikipedia).

Biomedical Model Organisms

The Fruit Fly (Drosophila melanogaster)

The "common fruit fly," shown in Fig. 6A.9, is widely used as a model organism for biological research in genetics, physiology, microbial pathogenesis, and life history evolution. In summer, attracted by "overripe fruit," those insects will look for the special smell released by the rotten fruit, flying in groups to the fruit to suck the remaining sweet substances or yeast. Hence Drosophila is also named fruit fly.

Figure 6A.9   Common fruit fly (source: Wikipedia).

Fruit flies can be said to be the best friends of geneticists. They can be readily reared in the laboratory, breed quickly, and lay many eggs. Female flies lay about 400 eggs each time, and in less than one day, the larvae can break out of the shell. They can pass 30 generations a year, plus they have only four pairs of chromosomes that are quite easy to analyze.

The genomic sequences of at least 12 species of Drosophila have been published, including *Drosophila melanogaster, D. pseudoobscura, D. sechellia, D. simulans, D. yakuba, D. erecta, D. ananassae, D. persimilis, D. willistoni, D. mojavensis, D. virilis,* and *D. grimshawi.* Like humans, fruit flies also originate in Africa. Between 10,000–15,000 years ago, they migrated from Africa to the rest of the world, along with humans.

The most popular fruit fly among scientists is *Drosophila melanogaster,* named after the black color of the male abdomen. After successful sequencing of the *Drosophila melanogaster* genome in 2000, it was found that the genome of *Drosophila melanogaster* is about 140 Mb in size. About 13,800 coding genes have been annotated, of which only about 77% are shared across the 12 published species listed above. Different regions in the Drosophila genome evolve at different rates. The genes involved in fruit fly taste, smell, detoxification, metabolism, reproduction, and immunity, develop the fastest. For example, *D. sechellia,* named after the Indian Ocean islands, has a single source of food, and the rate of gene loss associated with taste is five times that of other species. Drosophila and human genome sequence homology is surprisingly high, around 60%. As many as 75% of the known pathogenic genes in humans are similar to those in Drosophila, so their genes located on relatively simple chromosomes are suitable for human genetics and even disease mechanism research.

It is remarkable how many Nobel prizes have been awarded for research using Drosophila.

Our story must start with *Thomas Hunt Morgan,* whom we already met in Chapter 2. Morgan was originally a zoologist who studied embryos. In 1900, he encountered the work of Mendel. However, he decided to choose fruit flies—not peas—as his new experimental objects. In May 1910, Morgan discovered a mutant fly with white eyes (the fruit fly usually has red eyes). He cultivated various offspring in different ways and found that the ratio of red eyes to white eyes was 3:1. Further studies found that the offspring of white-eye males and red-eye females were all red flies. The male offspring of white-eye females and red-eye males all had white eyes and the females had red eyes. He came up with a theory of complex genes, and divided the eyes of the fruit fly, the shape of the wings, the appearance of the fluff, etc., and then cultivated these genetic characteristics. He studied the phenomenon of linkage between genes located on the same chromosome. Genetic linkage is the tendency of DNA sequences that are close together on a chromosome to be inherited together during sexual reproduction. Two genetic markers that are physically near to each other are unlikely to be separated during chromosomal crossover, and are therefore said to be more linked than markers that are far apart. In other words, the nearer two genes are on a chromosome, the lower the chance of recombination between them, and the more likely they are to be inherited together. On the other hand, markers on *different* chromosomes are perfectly "unlinked." Morgan established the genetic *law of linkage and crossing-over*. As suggested by English geneticist J. B. S. Haldane, the measurement unit for linkage is called the *morgan*. A centimorgan (abbreviated cM) or map unit (m.u.) is defined as the distance between chromosome positions (also termed loci or markers) for which the expected average number of intervening chromosomal crossovers in a single generation is 0.01. The number of DNA base-pairs to which it corresponds varies widely across the genome because different regions of a chromosome have different propensities towards crossover. In humans, one centimorgan corresponds—on average— to about 1 million base pairs. In 1913, Morgan's student Alfred Sturtevant developed the first genetic map. The fruit fly has proved the genetic chromosomal theory that genetic characteristics often are not affected by just one gene but by multiple genes. In 1933, Morgan received the Nobel Prize[110] "for his discoveries concerning the role played by the chromosome in heredity." T. H. Morgan became the first geneticist to be awarded.

In 1916, Thomas Hunt Morgan's student *Hermann Joseph Muller* discovered the genetic phenomenon of cross-interference in the fruit fly experiment. In July 1927, Muller published a research paper entitled "Artificial Transmutation of the Gene" in Science,[111] reporting the discovery of X-ray-induced gene mutations. The role of X-rays in inducing mutations was first confirmed, and the relationship between mutagens dose and mutation rate was clarified, thereby laying a theoretical foundation for mutagenesis breeding. The study had gone on for over 10 years from beginning to end, and the substantial breakthrough in the artificial mutagenesis experiment also earned him the Nobel Prize in 1946,[112] the second fruit fly-related award.

It is worth mentioning that in 1941, another student of Morgan, George Beadle, gave up two years of research on fruit fly genetics, working instead with biochemist Edward Tatum. They subjected red *Bacillus subtilis* (Streptomyces sp.) to X-ray irradiation and established proof that biochemical processes in all organisms are ultimately controlled by genes. Their research suggested the famous "one gene and one enzyme" theory. Both were rewarded with the 1958 Nobel Prize[113] in physiology or medicine, "for their discovery that genes act by regulating definite chemical events." Although they were not directly rewarded for fruit fly research, the scientific training and the research inspiration they received were related to fruit flies.

One of the most significant fruit fly related awards was the 1995 Prize won by Edward B. Lewis, Christiane Nüsslein-Volhard, and Eric F. Wieschaus for their discoveries concerning "the genetic control of early embryonic development."[114]

The newly laid egg of a fruit fly develops in 10 days, first to a larva, then to a pupa, and finally to a sexually mature fly (Fig. 6A.10). The Nobel Laureates identified and classified 15 genes of key importance in determining the body plan and the formation of body segments of *Drosophila melanogaster*. The adult fruit fly consists of head, 3 thoracic segments, 8 or 9 abdominal segments, and tail. The individual segments develop differently during embryogenesis. After having tested more than half of the approximately 14,000 genes of the fly, Nüsslein-Volhard and Wieschaus found three groups of genes that govern segmentation. The first group of genes provides a basis for segmentation along the body axis. These *gap-genes* are busy at the start, specifying a rough body plan. Loss of a gap gene results in a reduced number of segments.

The next group of genes, the *pair rule-genes*, govern formation of every second body segment. Loss of a pair rule-gene, e.g., even-skipped, allows only odd-numbered segments to develop.

Finally, the third group of genes, the *segment polarity-genes*, refine the structure of the individual segments, in particular head-to-tail polarity: they ensure that the head end and the tail end of a segment look different. Loss of a segment polarity-gene leads to drosophila body segments with similar head and tail ends.

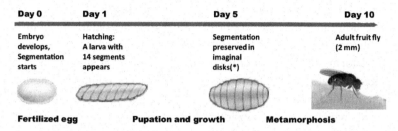

Day 0	Day 1	Day 5	Day 10
Embryo develops, Segmentation starts	Hatching: A larva with 14 segments appears	Segmentation preserved in imaginal disks(*)	Adult fruit fly (2 mm)

Fertilized egg   **Pupation and growth**   **Metamorphosis**

(*) An imaginal disk is one of the parts of an insect larva that will be part of the outside of the adult insect during the pupal transformation.

Figure 6A.10   Embryonic development from egg to adult fruit fly (adapted from Nobelprize.org).

Surprisingly, for the fruit fly there were only 15 genes responsible for embryonic development. Edward Lewis subsequently found that these genes were arranged one after another in the DNA of the fly. The order of the genes in the DNA also corresponded to the order of Drosophila's larval segments.

Those insights have paved the way for other developmental biologists to make pioneering discoveries across many other species. We humans have genes that are closely related to the genes discovered by the laureates, and they perform similar functions in our embryonic development.

In 2004, Richard Axel and Linda B. Buck of Columbia University, NY, earned the Nobel Prize[115] for their work on "odorant receptors and the organization of the olfactory system," which was based on Drosophila research. They discovered that Drosophila has a specific brain region in its olfactory function, clarified how the olfactory system works and found that quite a large fraction of our genome, containing about 1000 different genes, is focused on olfactory receptors. These receptors are located on the olfactory receptor cells, which occupy a small area in the upper part of the nasal epithelium and detect the inhaled odorant molecules. Drosophila research also provides a model for further elucidation of the relationship between genes, nerves (brains) and behavior.

In 2009, the Nobel Prize in physiology or medicine was awarded jointly to Elizabeth H. Blackburn, Carol W. Greider, and Jack W.

Szostak, for the discovery of "how chromosomes are protected by telomeres and the enzyme telomerase."[116] They discovered how the chromosomes can be copied in a complete way during cell divisions and how they are protected against degradation by means of telomeres (the ends of the chromosomes) and by the enzyme that forms them: telomerase. If the telomeres are shortened, cells age. Conversely, if telomerase activity is high, telomere length is maintained, and cellular death is delayed. This is the case in cancer cells, which can be considered to have "eternal life." Although involving chromosomes, the model organisms used were mostly yeast and *C. elegans*, rather than the fruit fly.

The 2011 Nobel Prize in physiology or medicine[117] was shared by Bruce A. Beutler, Jules A. Hoffmann, and Ralph M. Steinman, for discoveries related to our understanding of the immune system, by discovering key principles for its activation. Among them, Jules Hoffman and his colleagues studied how fruit flies fight infection in 1996. Bruce Beutler was looking for a receptor that binds to bacterial products. In addition to doing fruit fly experiments, he also did mouse experiments.

In 2017, Jeffrey Hall, Michael Rothsbach, and Michael Young won the Nobel Prize[118] for revealing the genetic code of the human circadian clock mechanism, discovered in fruit flies. Research on genes related to circadian clocks already started in the 1960s and 1970s. However, it wasn't until the mid-1980s that these three U.S. scientists isolated a gene from Drosophila that could control the circadian rhythm. Actually, the first appearance of the circadian rhythm in the history of science begins with a pot of smart mimosa, as shown in Fig. 6A.11. In 1729, the French geophysicist and astronomer Jean-Jacques d'Ortous de Mairan observed that mimosa would unfold in the spirit during the day, and that the leaves would close and hang down at night, as if the sun gave it special guidance. Mimosa maintains a stable rhythm whether it is shaded or not. In 1823, the Swiss botanist Augustin de Candolle, regained the study, placing mimosa in a state of light and heat. He then measured that it spontaneously showed a work cycle of about 22–23 hours. This may have been the first time in history that humans have realized that there is a clock-like inner rhythm in the living body.

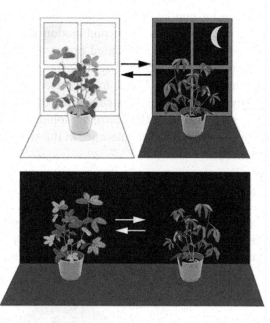

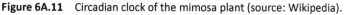

Figure 6A.11 Circadian clock of the mimosa plant (source: Wikipedia).

Later, successive studies have found that there are similar biological clocks in various animals and plants. But research on the circadian clock only stayed at the organ or tissue level, the deeper mechanisms were still not known. It wasn't until 1971 that Seymour Benzer and his student Ronald Konopka experimented on fruit flies to discover a gene that could affect the circadian rhythm. The new gene, named "period," acts as the "cycle" gene. However, how does the cycling actually work? This question was answered by the three Nobel Prize winners in 2017.

To show how the fruit fly is able to adapt to extreme conditions, consider experiments where Drosophila was exposed to extremely low (hypoxic) oxygen levels, equivalent to altitudes far above Earth's highest mountain, Chomolangma (Everest): Experiments carried out in the lab of Prof. G. Haddad,[119] UC San Diego, have shown that Drosophila may even exceed the naked mole rat's (see separate chapter below) resistance to hypoxia: by the 13th generation, fruit flies were able to complete development and perpetually live in 5% O_2; and by the 32nd generation, the hypoxia selected flies could even live perpetually under an even more severe level (4% of O_2), a lethal condition for non-adapted fruit flies. This is probably due to newly occurring mutations or recombination and selection of favorable alleles in the hypoxia adapted population. In fact, after 8 generations in normoxia, when those hypoxia selected fruit flies were re-introduced into the lethal hypoxic environment of 4% O_2, the majority (>80%) of the flies completed their development and could be maintained in this extreme condition perpetually. This result suggests that hypoxia tolerance in the selected flies is a heritable trait.

The Mouse (Mus musculus)

A mouse is a small rodent characteristically having a pointed snout, small rounded ears, a body-length tail and a high breeding rate. The most common mouse species is the house mouse, shown in Fig. 6A.12.

Figure 6A.12   House mouse (source: Wikipedia).

Although preyed upon by cats, wild dogs, foxes, birds of prey, snakes, etc., its remarkable adaptability to almost any environment has made the mouse one of the most successful mammals living on Earth today.

Mice are commonly used as mammalian "model organisms" in biological and medical research, because they share a high degree of genetic homology (shared ancestry) with humans. Virtually all mouse genes have human homologs. The mouse has approximately 2.7 billion base pairs and 20 chromosomes.

There is no animal that has been involved in more research projects leading to Nobel Prize awards (try to search for "mouse" on nobelprize.org and you will get close to 500 hits!), and there is no animal more popular as a cartoon or a stuffed animal than Disney's Mickey Mouse.

Reasons for the common selection of mice as model organisms are their small size, low cost and widely varied diet, easy maintenance, and fast reproduction rates. The average gestation period of mice is 20 days, and pups are weaned at 3 weeks of age. Therefore, several generations of mice can be observed in a relatively short time. Mice are generally very docile if raised from birth and given sufficient human contact.

The history of the mouse as model organism for research is closely coupled to the history of the Jackson Laboratory:

The Roscoe B. Jackson Memorial Laboratory[120] was founded in 1929 by Clarence Cook Little, a Harvard-trained geneticist who had served as president of both the Universities of Maine and Michigan. Little created the first inbred (genetically uniform) mouse strains and wanted to use them to establish that cancer was a genetic disease, not an infectious disorder, as was widely thought at the time. He gained initial financial support for the laboratory from Detroit industrialists Edsel Ford (president of the Ford Motor Company) and Roscoe B. Jackson (president of the Hudson Motorcar Company). The lab was named (in 1963) after Jackson, and Bar Harbor, Maine, was chosen as its site, built on land donated by Little's family friend, George B. Dorr.

In 1933 the laboratory began providing genetically defined mouse strains to the scientific community as experimental models for various human diseases.

The first edition of *Biology of the Laboratory Mouse*, the first book devoted to mouse biology and genetics, was published in 1941.

In 1974, the Mouse Genetics Laboratory was dedicated. In 1976, Jackson Lab had grown to 450 employees, housed 700,000 mice, and was the world's largest mammalian genetics research center.

In 1983, NIH's National Cancer Institute (NCI) awarded a grant that made the Jackson Laboratory the only mammalian laboratory designated as a Cancer Center.

In 2000, the Jackson Lab extended its work on mouse models for human neurological diseases such as epilepsy, addiction and neurodegenerative disorders. The same year it established a center for mouse models of heart, lung, blood and sleep disorders. Later on, obesity and diabetes research models were added.

Today's mission of the JAX Lab is stated as follows: "More than 1,800 employees are working toward the discovery of precise genomic solutions for disease, and to empower the global biomedical community in the shared quest to improve human health." Although there is an important difference between mouse biology and human biology, there is no doubt that Little's original idea, to study the mouse-man connection, has been highly fruitful.

Finally, here are (probably still incomplete) statistical data related to Nobel Prizes awarded to scientists using mouse models for biomedical and chemical research:

During the 60 years from 1906 to 1965, 10 Nobel Prizes were earned for enhancing our knowledge about neurons, the adaptive and innate immune system, infectious diseases and first attempts to cure and prevent them, and early attempts to extract insulin from the pancreas of domestic animals.

During the 30 years from 1966 to 1995, 16 Nobel Prizes were awarded for using laboratory mice to achieve breakthroughs in cell biology, tumor biology, the communication between neurons, molecular genetics, transplantation immunology, the origin and dissemination of infectious diseases, and the genetic control of embryonic development.

Finally, during the 21 years from 1996 to 2016, as many as 21 Nobel Prizes were based on further research activities involving mouse models.

In fact, it would probably be a highly worthwhile and interesting task to write a book about the close to 50 Nobel Prizes involving *Mus musculus*.

Let's just pick two very important fields of biomedical research, namely monoclonal antibodies and embryonic stem cells, both highly dependent on mice.

To explain the concept of *monoclonal antibodies*, let's go back to the work by Paul Ehrlich,[121] who shared the 1908 Nobel Prize in physiology or medicine for his research into "immunity." Ehrlich not only developed chemotherapy but is most famous for introducing the idea of "magic bullets," substances that could selectively target organisms that caused diseases such as sleeping sickness, typhoid, and syphilis. We know that our body's immune system is making "antibodies," proteins that recognize invading microbes and other attacking organisms and then try to destroy the dangerous targets. Monoclonal antibodies (mab) are defined as "monospecific antibodies that are made by identical immune cells that are all clones of a unique parent cell." Following Paul Ehrlich's magic bullet idea, the goal is to produce monoclonal antibodies that specifically bind to a given substance and can then serve to detect or purify that substance.

In 1984, Georges Köhler, César Milstein, and Niels Kaj Jerne shared the Nobel Prize in physiology or medicine for "theories concerning the specificity in development and control of the immune system and the discovery of the principle for production of monoclonal antibodies."[122]

Milstein and Köhler were able to turn the normally evil feature of tumor cells, the capacity to proliferate forever, into a very beneficial property: Using a technique to fish up antibody-producing cells from a sea of cells, they then fused these cells with mouse tumor cells and created hybrid cells with a capacity to produce the very same antibody in high quantity. They called these hybrid cells hybridomas. Since all cells in a given hybridoma come from one single hybrid cell, the antibodies made are *monoclonal.* The first application of "hybridoma technology" was to use a line of myeloma cancer cells and to fuse them with healthy antibody-producing B cells. Monoclonal antibody (mab) technology allowed scientists to grow huge quantities of pure antibodies aimed at specific selected targets, leading to the design of new diagnostic tests and therapeutics. By injecting a payload of mabs into the bloodstream, the antibodies were headed straight to their disease target.

Drugs based on monoclonal antibody technology (all named xxx-mab) are now among the most important and best-selling drugs. We will discuss them further below in our chapter on genetic engineering.

Although it took more than 20 years from Nobel Prize-winning research (involving mice) to first therapeutic and commercial success, monoclonal antibodies have now been generated and approved to treat cancer, cardiovascular disease, inflammatory diseases, macular degeneration, transplant rejection, multiple sclerosis, and viral infection.

Another highly significant mouse-related Nobel Prize was awarded in 2007[123] jointly to Mario R. Capecchi, Martin J. Evans, and Oliver Smithies, for their discoveries of "principles for introducing specific *gene modifications in mice by the use of embryonic stem cells.*" Their breakthrough research, carried out since the 1980s, marked the beginning of a new era in genetics and led to an explosion of biomedical research activities based on the role of genes involved in mammalian organ development, and based on the development of mouse models for human diseases.

Gene targeting is often used to inactivate single genes. Such gene "knockout" experiments have elucidated the roles of numerous genes in embryonic development, adult physiology, aging, and disease. To date, most mouse genes (approximately half of the genes in the mammalian genome) have been knocked out. "Knockout mice" are now routinely created and distributed by institutions such as the Jackson Lab.

With gene targeting it is now possible to produce almost any type of DNA modification in the mouse genome, allowing scientists to establish the roles of individual genes in health and disease. Gene targeting has produced hundreds of different mouse models of human disorders, including cardiovascular and neurodegenerative diseases, diabetes, and cancer.

To understand how knockout mice are created, we need to go back to the role played by DNA in carrying information about the development and function of our bodies throughout life. Our DNA is packaged in chromosomes, which occur in pairs—one inherited from the father and the other from the mother. Exchange of DNA sequences within such chromosome pairs increases genetic variation in the population and occurs by a process called *homologous recombination*. Homologous recombination is a type of genetic recombination in which nucleotide sequences are exchanged between two similar or identical molecules of DNA. It is used by our cells to accurately repair harmful breaks that occur on both strands of DNA. Homologous recombination also produces new combinations of DNA sequences during meiosis, the process by which eukaryotes make gamete cells, like sperm and egg cells in animals. These new combinations of DNA represent genetic variation in offspring, which in turn enables populations to adapt during the course of evolution. The 2007 Nobel Laureates Mario Capecchi and Oliver Smithies first demonstrated that homologous recombination could be used to specifically modify genes in mammalian cells.

However, the cell types initially studied by Capecchi and Smithies could not be used to create gene-targeted animals. This required another type of cell, one which could give rise to germ cells, thereby enabling inheritance of DNA modifications.

That second part of the solution (Fig. 6A.13) was provided by Martin Evans, who had the vision to use mouse embryonal carcinoma (EC) cells as vehicles to introduce genetic material into the mouse germ line. His attempts were initially unsuccessful because EC cells carried abnormal chromosomes and could not therefore contribute to germ cell formation. Evans then discovered that chromosomally normal cell cultures could be established directly from early mouse embryos. These cells are now referred to as *embryonic stem* (ES) cells.

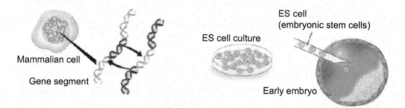

Figure 6A.13 Introduction of specific gene modifications in mice requires both homologous recombination (left picture) and use of embryonic stem cells (right picture) to ensure inheritance (adapted from Nobelprize.org).

To demonstrate that ES cells can be genetically modified and still continue to multiply and give rise to a new organism, Evans injected mouse embryos with ES cells that previously had been injected with DNA from a virus. The embryos were implanted in the uterus of a surrogate mother where they developed just like normal baby mice. When they were born it was possible to find viral DNA in the genome of some of the mouse pups, demonstrating that inserted DNA could in fact be passed down (see Fig. 6A.14).

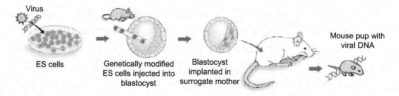

Figure 6A.14 Using embryonic stem cells to pass inserted DNA to mouse pups (adapted from Nobelprize.org).

The Nobel Committee awarded the 2007 Prize to Capecchi and Smithies, and to Evans, because it was the combination of their respective ideas that enabled the subsequent creation of mouse models.

And so the mouse continues to be the most useful animal in biomedical research.

Let's review in detail what we know about mouse genomics. About 80% of human and mouse genes correspond to each other. The human genome is longer, about 3 billion base pairs on 23 chromosomes, compared to 2.5 billion base pairs on 20 chromosomes for the mouse, and the types of genes are also somewhat different. For example, mice have more reproductive, immune, and olfactory genes than humans. To study differences in embryonic development, Settembre et al. compared the gene sequence of mice with human chromosome 21, and tracked the expression of about 160 genes involved in mouse embryos at different stages of development.[124] The scientists monitored which genes are still being expressed in organs of adult mice (and rats), such as the brain, muscle, heart, etc., and used this method to identify genes of interest, including the Adarb1 gene that plays a role in cardiac development.

From a genetic point of view, there are several special reasons why mice are so useful and valuable for human biomedical research.

1. The mouse genome is highly homologous to the human genome. From an evolutionary point of view, the mouse as a representative of mammals has its unshakable position. In addition to mice, four other animal models (zebrafish, Xenopus,[125] fruit flies, and nematodes) have evolved differently from humans at least 270 million years ago, and mice 60 million years ago. It shares an ancestor with mankind. In 2002, when the sequencing of the mouse genome was initially completed,[126] it was found that 96 of the first analyzed 99 mouse genes were homologous to human genome sequences. It was thus confirmed that mice and small rats are highly homologous to humans at the genetic level. The fragments between genes and genes on the DNA strand are also very similar. Only a few hundred genes are unique to a particular species; therefore, information on human diseases and physiological functions can be obtained

through comparative studies. The physiological, biochemical, and regulatory mechanisms of mice are the same or similar to those of humans.

2. The experimental results obtained using inbred mice with genetic backgrounds that are essentially identical are easily replicated in different laboratories, which makes the experimental results more readily acceptable.

3. The successful establishment of a mouse ES cell line made it possible for humans to modify the mouse genome fairly easily, and the development of transgenes, gene knockouts, and gene knock-in techniques resulted in a large number of genetically modified mouse models. When the Human Genome Project was nearing completion, the idea of mutating all mouse protein-coding genes was proposed by scientists at Cold Spring Harbor in 2003. Recently, scientists have discussed the massive humanization of mouse genes to generate many more human-like disease models. This may be another major strategic direction following the large-scale mouse gene knockout program. In addition, a series of new technologies (such as TALEN[127] and CRISPR/Cas9 technologies) that have recently been established, reduced the difficulty of developing genetically engineered mice, increased their efficiency, and reduced costs. More about CRISPR in Chapter 7.

Here is some recent news about the mouse as a model organism to help us understand the molecular basis of hearing (across all vertebrates, including fish, amphibians, reptiles, and mammals, including humans). On August 22, 2018, a Harvard Medical research team led by David Corey and Jeffrey Holt ended a 40-year quest for the identity of the sensory protein responsible for hearing and balance.[128] The transmembrane channel-like protein (TMC) 1, associated with the TMC1 gene which was discovered in 2002, is the critical molecular sensor that converts sound and motion into electrical signals that the brain can process. TMC1 is the "gatekeeper of hearing." Considering that, worldwide, an estimated 460 million people[129] suffer from loss of hearing, making this condition the #1 neurological disorder, this new molecular insight could open the door to new medical treatments. By using a combination of computer simulations and experiments in the lining hair cells of mice, the scientists found that TMC1 proteins assemble in pairs to form "sound-activated pores" that open ion channels for signal transmission. Hearing loss occurs when the TMC1 molecular gate is malformed or missing. Already in 2015, it was found that genetically deaf mice treated with TMC1 gene therapy recovered some of their hearing.[130]

As a curiosity, it may finally be worth mentioning that at least three of the aforementioned "mouse hits" in the Nobel Foundation website involve recipients of the Literature Prize. In 1962, John Steinbeck gave a memorable acceptance speech where he talked about "not squeaking like a grateful and apologetic mouse, but to roar like a lion."

Instead, humans should be grateful. We owe a lot to our squeaky mouse companions despite the widespread view that they are nothing but a nuisance to Homo sapiens.

California Sea Slug (Aplysia californica)

Sea slug is a common name for some sea snails (marine gastropod mollusks) that over evolutionary time have seemingly lost their shells. Sea slugs have enormous variation in body shape, color, and size. Most are partially translucent and have bright colors. These colors attract predators but may also serve as a warning to other animals of the sea slug's poisonous stinging cells.

The California sea slug (Fig. 6A.15) has been made famous by neuroscientist Eric Kandel (Nobel Prize 2000[131]), who used it as laboratory animal in his studies of the neurobiology of learning and memory. Kandel was born in Vienna, Austria, but his family escaped to New York when Hitler came to power and started to persecute Jews. The traumatic experience when Gestapo invaded his apartment on his birthday has stayed with him all his life, and has been described in his award-winning scientific memoir *In Search of Memory*.[132] This chapter is very much based on his writing.

What makes Aplysia so special is its simple nervous system, which consists of just 20,000 large and easily identified neurons with cell bodies up to 1 mm in size. By studying the sea slug, Kandel realized that memory storage must rely on modifications in the synaptic connections between neurons. Earlier comparative studies of animal behavior by 1973 Nobel Laureates[133] Konrad Lorenz ("imprinting of birds"), Niko Tinbergen ("instinctive bird behavior"), and Karl von Frisch ("language of bees"), had already revealed that simple forms of learning could also be found in animals. Kandel felt that an animal model could facilitate electrophysiological analysis of the synaptic changes involved in learning and memory storage. It turned out that, ultimately, his results would be found to be applicable to humans. Although many senior biologists and psychologists believed that nothing useful could be learned about human memory by studying the physiology of simple animals, his results proved his doubters wrong: The human brain may have about 100 million times more nerve cells than a sea slug (as illustrated in Fig. 6A.16), but the basic mechanism of memory storage is chemically equivalent in both species.

Figure 6A.15 *Aplysia californica* releasing ink after being disturbed (source: Wikipedia).

The sea slug has a withdrawal reflex protecting its gills.[134] If the gills are touched repeatedly, they react less and less. The short-term *habituation* or *amplification* effect lasts only for a few minutes and can be classified as "short-term memory." However, if the touch is forceful the reflex is amplified and the sea slug's reaction becomes stronger and stronger.

By studying the Aplysia gill-withdrawing reflex, Kandel found that, prior to training, a stimulus to a sensory neuron in Aplysia might be strong enough to cause motor neurons leading to the gill to fire action potentials, but not strong enough to do the same for motor neurons leading to the ink gland. Subsequent *training* causes neurons to grow new terminals, thereby strengthening the synapses between the sensory neuron and the motor neurons of *both* the gill and the ink gland. After training, when the sensory neuron is stimulated, both motor neurons will fire action potentials, causing gill withdrawal *and* inking. Kandel's experiment proved that *learning* gives rise to a physical change in the brain of Aplysia.

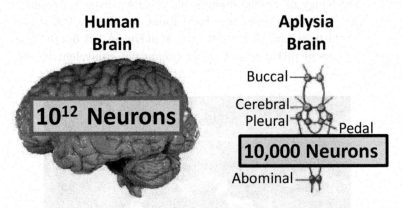

Figure 6A.16 There are only 20,000 Aplysia nerve cells compared to 10^{12} human neurons (included by permission of Prof. E. Kandel and Nobelprize.org).

By studying Aplysia in vivo, Kandel obtained a full understanding of the underlying biochemical processes responsible for short-term memory. Cyclic AMP (adenosine monophosphate) and protein kinase A are both necessary for strengthening the connections between sensory and motor neurons. Serotonin is responsible for turning on the signaling, and the cell signaling uses the potassium ion channel.

The strong stimulus that forms the long-term memory works in a completely different way compared to short-term memory. To find out how, Kandel could no longer just conduct in vivo studies of Aplysia. The study of long-term memory required measurements of neural populations over several days. Removing sensory and motor cells from adult animals did not work because adult cells do not survive well in culture. Kandel and his students therefore learned how to grow Aplysia in the laboratory. They learned about the importance of red seaweed in transforming larvae into juvenile slugs and soon were able to extract sensory neurons, motor neurons and (serotonin releasing) interneurons involved in the gill withdrawal reflex. They had assembled all the elements of a learning circuit in tissue culture and were ready to conduct *in vitro* experiments, based on synaptic connections equivalent to the "intact" Aplysia animal.

Being fully aware of the insights gained by Jacob and Monod (see Chapter 2), Kandel and his team paid special attention to gene expression in the neurons under investigation. They formed the hypothesis that long-term memory would require anatomical changes, in particular synthesis of new proteins within the cell nucleus. While single pulses of serotonin increase cyclic AMP and kinase A primarily at the synapse, repeated strong pulses of serotonin cause protein kinase A to move into the cell nucleus where it first recruits another kinase, called MAP kinase, then activates a new "cyclic AMP response element-binding protein (CREB)" which then binds to a "promoter" gene, the cyclic AMP response element. It turns out that CREB is a key component of the switch that converts short-term synaptic connections to long-term facilitation and the growth of new synaptic connections. In other words, an environmental stimulus (shock to animal's tail) activates serotonin which then acts on sensory neurons to increase cyclic AMP, then causes protein kinase A and MAP kinase to move to the nucleus and activate CREB, which then causes genes to be expressed at the synapse that produce a long-term change.

It was found later that there are two forms of CREB, namely CREB-1 which activates gene expression and leads to long-term memory, and another form (CREB-2) that suppresses gene expression and inactivates long-term memory. It was further found that MAP kinase inactivates CREB-2 and that protein kinase A activates CREB-1. The question then arises why there is a need for CREB-2? The answer probably is that not all long-term memories are worth being stored forever, that there should be a mechanism for erasing memories.

There is a final piece to the puzzle of ("implicit") long-term memory—although we should be very careful to use the word "final" when it comes to biology: Nothing seems ever to be just "true" or "false," nothing ever seems to be *fully* understood! Anyway, for Kandel the question remained about a possible role played by localized proteins, proteins present at some (but not all) synapses attached to a neuron. It turned out that a necessary requirement for SUSTAINED long-term memory is the presence of a localized protein known as cytoplasmic polyadenylation element-binding (CPEB) protein. A sensory neuron may have some 1200 synaptic terminals and it makes contact with about 25 target motor neurons. However, when CREB regulator genes send messenger RNA to the synapses, only the ones "marked for growth" by means of protein kinase A will be activated and will form local CPEB proteins. An important feature of CPEB is that it is "self-perpetuating," i.e., it maintains local protein synthesis at the synaptic terminals marked by strong physiological serotonin signals. Those CPEB proteins take over the job to sustain the long-term synaptic connection as long as it's not turned off by CREB-2 action. Figure 6A.17 illustrates what Kandel calls a "dialog between genes and synapses."

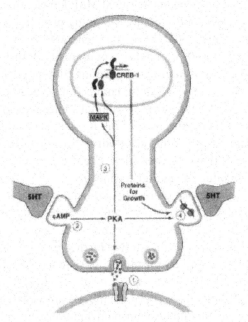

Figure 6A.17 (1) Neurotransmitter (serotonin in Aplysia) activates ion channel in a sensory neuron and leads to synaptic action. (2) Cyclic AMP and kinase A are activated, leading to "short-term" synaptic connection with motor neuron. (3) Repeated strong serotonin activation leads to translocation of the kinase to the nucleus and to activation of gene transcription. (4) Activation of local protein synthesis at the synapse stabilizes the long-term synaptic connection. 5HT stands for serotonin, and PKA for protein kinase A (included by permission of Prof. E. Kandel and Nobelprize.org).

It is probably also true that people known to have good memory may have a genetic bias towards CREB-1 and are limiting the activity of CREB-2. Full elimination of CREB-2, however, could rather be a curse! Who wants to have all childhood and other memories stay in your mind forever?

An important lesson learned by Kandel's pioneering work based on Aplysia neurons is that genes are responsive to environmental simulation. The genes activated by CREB-1 are needed for new synaptic growth and the genes activated by CREB-2 proteins are making sure that we do not remember everything that we have ever learned. Those CREB-2 activated suppressor genes encode suppressor proteins that will erase synaptic connections and hence memories.

What we call memory in Aplysia and in most other animals including sapiens, is thus elicited by direct physical changes in the billions of synapses that form the contact points between the nerve cells. Note that Kandel's Aplysia based research explains what we now call "implicit memory." To take the next step and explain "explicit memory" where the mammalian brain participates extensively, he had to move from Aplysia to the mouse as model organism. When doing so, brain scientists later learned that long-term memories associated with strong emotions are much more difficult to erase

Because human mental processes had long been thought to be unique, Kandel met resistance when he tried to transfer his Aplysia based biochemical learnings to mammalian and, ultimately, human brains. However, there are very few proteins that are truly unique to the human brain, and the signaling systems are the same across living organisms.

"All life, including the substrate of our thoughts and memories, is composed of the same building blocks."[136]

The Zebrafish (Danio rerio)

The use of zebrafish as a laboratory animal (in addition to its popularity as aquarium fish) was pioneered by molecular biologist George Streisinger and his colleagues at the University of Oregon in the 1970s and 1980s. Figure 6A.18 shows a zebrafish along with a picture of Streisinger.

Figure 6A.18 The zebrafish is a model organism, forever connected with George Streisinger (source: Wikipedia).

Streisinger's life story is similar to Eric Kandel's (see the Sea Slug chapter). Born in Budapest in 1927, his Jewish family left Hungary in 1937 for New York to escape Nazi persecution. Streisinger graduated from the highly regarded Bronx High School of Science in 1944, and then went on to study at Cornell University, at the University of Illinois, and the California Institute of Technology in Pasadena, Southern California. He had the great fortune of being mentored by Theodosius Dobzhansky, an outstanding geneticist and evolutionary biologist, professor at NY Columbia University, and author of the book *Genetics and the Origin of Species* (1937).[137]

In 1960, Streisinger moved North to the University of Oregon, Institute of Molecular Biology, where he established a reputation as the "father of zebrafish research." In 1974, he applied for US government research funds by describing his goals as follows: "study features of the organization and embryological development of the vertebrate nervous system through the use of mutant strains." He was particularly interested in the mechanisms leading to the formation of specific synaptic connections and in the nature of the signals that guide specific axons to particular target sites.[138] In 1981, he published a breakthrough paper in *Nature*[139] about Zebrafish cloning.

The zebrafish is a freshwater fish that is native to the Himalayan region. It is particularly notable for its regenerative abilities.

Streisinger's zebrafish clones were among the earliest successful vertebrate clones created. Their importance has been consolidated by successful large-scale forward genetic screens (commonly referred to as the Tübingen/Boston screens). The fish has a dedicated online database of genetic information, the Zebrafish Information Network

(ZFIN),[140] which is managed by the University of Oregon. ZFIN is a web-based community resource that serves as a centralized location for the curation and integration of zebrafish genetic, genomic and developmental data. Similarly, the Zebrafish International Resource Center (ZIRC),[141] another genetic resource managed by University of Oregon, is making close to 39,000 zebrafish lines available for distribution to the research community. The Zebrafish book, dedicated to the memory of George Streisinger, guides users how to deal with Danio (previously Brachydanio) rerio in the laboratory.

As an aquarium fish, zebrafish are hardy fish with a playful disposition. They are nice to watch, cheap, and broadly available. They also interact well with other fish species in the aquarium. In captivity, zebrafish can live approximately 3.5 years. In late 2003, transgenic zebrafish that express green, red, and yellow fluorescent proteins became commercially available in the United States. The fluorescent strains are trade-named GloFish. The leopard danio, previously known as Danio frankei, is a spotted color morph of the zebrafish which arose due to a pigment mutation. Other forms of both the zebra and leopard pattern, along with long-finned subspecies, have been obtained via selective breeding programs for the aquarium trade.[142] Various transgenic and mutant strains of zebrafish were stored at the China Zebrafish Resource Center (CZRC),[143] a nonprofit organization, supported by the Chinese Academy of Sciences.

In 1999, the *nacre* mutation was identified in the zebrafish ortholog of the mammalian MITF transcription factor. Human MITF mutations result in eye defects and loss of pigment. The zebrafish gene responsible for its unusual pigmentation was identified as SLC24A5, a solute carrier required for melanin production. The orthologous human gene was then found to strongly segregate fair-skinned Europeans and dark-skinned Africans.[144] Zebrafish with the nacre mutation have since been bred to make fish that are transparent into adulthood. These fish are characterized by uniformly pigmented eyes and translucent skin.

In 2008, researchers at Boston Children's Hospital developed a new strain of zebrafish, named Casper, whose adult bodies had transparent skin.[145] This allows for detailed visualization of cellular activity, circulation, metastasis, etc. The Casper strain is expected to yield insights into human diseases such as leukemia and other cancers.

As a model biological system, the zebrafish genome has attracted the interest of several leading research institutions. The Sanger Institute[146] (UK) started zebrafish genome sequencing in 2001, and the full genome sequence of the Tübingen[147]reference strain is available at the NIH National Center for Biotechnology Information (NCBI)'s Zebrafish Genome Page. The zebrafish reference genome sequence is annotated as part of the Ensembl project, and is maintained by NCBI's Genome Reference Consortium.[148] In 2009, the Institute of Genomics and Integrative Biology in Delhi, India, announced the sequencing of the genome of a wild zebrafish strain, containing an estimated 1.7 billion base pairs.[149] Comparative analysis with the zebrafish reference genome revealed a significant number of insertion deletion variations. Finally, the zebrafish reference genome sequence of 1.4 GB and over 26,000 protein coding genes was published in 2013.[150]

Although humans may appear to differ significantly from zebrafish, there are surprising similarities: 70% of human genes are found in zebrafish. Moreover, zebrafish have two eyes, a mouth, brain, spinal cord, intestine, pancreas, liver, bile ducts, kidney, esophagus, heart, ear, nose, muscle, blood, bone, cartilage, and teeth. Many of the genes and critical pathways that are required to grow these features are highly conserved between humans and zebrafish. Thus, any type of disease that causes changes in these body parts in humans could theoretically be modeled in zebrafish.[151]

The Zebrafish has a rapid embryonic development and its embryos are quite large, robust, and frequently transparent. Embryonic development can be readily observed because the embryos develop outside their mother.[152] Well-characterized mutant strains are readily available. The zebrafish is also demonstrably similar to mammalian models and humans in toxicity testing. It even exhibits a diurnal sleep cycle with similarities to mammalian sleep behavior.

Zebrafish have the amazing and exceptional ability to regenerate their heart cells and lateral line hair cells during their larval stages. Zebrafish eyes have also been found to regenerate photoreceptor cells and retinal neurons following injury. Scientists frequently amputate the zebrafish tail fins and analyze their regrowth to test for mutations: amputation switches zebrafish cells to an "active," regenerative, stem cell-like state. In 2012, Australian scientists published a study revealing that zebrafish use a specialized protein, known as fibroblast growth factor, to ensure that their spinal cords heal without glial scarring after injury.[153] Several important signaling pathways such as *Wnt* signaling[c] and Fibroblast growth factor have been found to be involved in the process.[154] Zebrafish research has also led to new insights into the complexities of human musculoskeletal diseases, such as muscular dystrophy.[155] Another focus of zebrafish research is to understand a gene called "hedgehog," a biological signal that underlies a number of human cancers and controls cell growth.[156]

Use of zebrafish in medical research

Zebrafish have been used to make several transgenic models of *cancer*, including melanoma, leukemia, pancreatic cancer, and hepatocellular carcinoma. The BRAF melanoma model was utilized as a platform for two screens published in March 2011 in the journal *Nature*. In one study, the model was used as a tool to understand the functional importance of genes known to be amplified and overexpressed in human melanoma.[157] One gene, SETDB1, markedly accelerated tumor formation in the zebrafish system, demonstrating its importance as a new melanoma oncogene. Another study revealed that an inhibition of the DHODH protein (by a small molecule called leflunomide) prevented development of the neural crest stem cells, which ultimately give rise to melanoma. Leflunomide may have utility in treating human melanoma.[158] It should also be noted that DHODH is a mitochondrial protein, and that inhibitors of this enzyme are used to treat autoimmune diseases such as rheumatoid arthritis.

In *cardiovascular* research, the zebrafish has been used to model blood clotting, blood vessel development, heart failure, and congenital heart and kidney disease.[159]

Immune system: In research into acute inflammation, scientists have established a zebrafish model of inflammation, and its resolution. This approach allows detailed study of the genetic controls of inflammation, including the possibility of identifying new

---

[c]The Wnt signaling pathways are a group of signal transduction pathways associated with proteins that pass signals into a cell through cell surface receptors. Wnt is an acronym that stands for "Wingless/Integrated."

drugs.[160] Zebrafish has also been used as a model organism to study vertebrate innate immunity.[161]

Infectious diseases: Many human infectious diseases can be modeled in zebrafish. The transparent early life stages are well suited for in vivo imaging and genetic dissection of host–pathogen interactions. For example, the zebrafish model for tuberculosis provides fundamental insights into the mechanisms of pathogenesis of mycobacteria.[162] Robotic technology has been developed for high-throughput antimicrobial drug screening using zebrafish infection models.

Repairing retinal damage: A notable characteristic of the zebrafish is that it possesses four types of cone cell, with ultraviolet-sensitive cells supplementing the red, green, and blue cone cell subtypes found in humans. Zebrafish can also be studied to better understand the development of the retina; in particular, how the cone cells of the retina become arranged into the so-called "cone mosaic." Zebrafish are known to have extreme precision of cone cell arrangement.[163] The zebrafish's retinal characteristics also show promise for human medicine: In 2007, scientists at University College London grew a type of zebrafish adult stem cell and hope to inject them into the eye to treat diseases that damage retinal neurons—including macular degeneration, glaucoma, and diabetes-related blindness. The stem cells successfully migrated into diseased rats' retinas, and took on the characteristics of the surrounding neurons. What remains is the challenge of translation of the approach to help humans.[164]

Muscular dystrophy **(MD)** is the collective term for a heterogeneous group of genetic disorders that cause muscle weakness and abnormal contractions, often leading to premature death. Zebrafish is widely used as model organism to study muscular dystrophies[165]. For example there is a zebrafish mutant that is a model organism for human Duchenne muscular dystrophy (DMD). Zebrafish is also an excellent animal model to study congenital muscular dystrophies, including CMD type 1 A (CMD 1A) caused by mutation in the human laminin $\alpha2$ (LAMA2) gene.[166] The zebrafish has become a model of choice in screening and testing new drugs against muscular dystrophies.

6B. Human Health

In this chapter we are covering a number of animals that either are linked already to the therapeutic treatments of human disease, or may have the potential to develop such a human connection.

A classic example is the extraction of insulin from the pancreas of pigs and cattle, to help patients suffering from diabetes. This practice went on until the manufacturing of human insulin by means of genetic engineering, at the very beginning of the biotech revolution.

Another example is how scientists have studied venoms secreted by all kinds of animals to develop drugs in therapeutic areas, including (so far) cardiovascular, pain, and type 2 diabetes medicine. As suggested in a recent *Science* paper by M. Holford et al.,[167] we should rely on the teachings of evolution to focus venom research on the discovery of human therapeutics and bioinsecticides.

Many animal species—not only snakes—have developed venomous capabilities, thereby moving predator–prey interactions from the physical to the biochemical domain. Venoms are typically used *both* to attack and to defend. Due to the ongoing "evolutionary arms race" between venomous animals and their prey, most venoms are quite complex and do exactly what biopharmaceutical drug discovery is aiming for: pursue molecular targets with high selectivity and potency! As to venom chemistry, it has been studied since mid-19th century, in particular with snakes. Proteins are by far the most important components of snake venoms (and other venoms as well). Among thousands of proteins found, there are toxins of various kinds with very specific properties:

Neurotoxins cause involuntary muscle contractions by disabling a mechanism that ends the signal at work in a synaptic connection (ion channel) between neurons. The target motor neuron is staying active when it no longer should, leading to sometimes fatal consequences for the prey. Neurotoxins are also used by other venomous organisms such as sponges, spiders, scorpions, bugs, wasps, etc.

Dendrotoxins inhibit neurotransmissions by blocking the exchange of positive and negative ions across the neuronal membrane. The effect is paralysis of the nerves of the prey. A similar result is generated by *α-neurotoxins*, a large (over 100) group of proteins that cause numbness and paralysis by blocking the flow of the neurotransmitter acetylcholine (ACh).

Cytotoxins destroy certain cell mechanisms: *Phospholipase* is an enzyme that transforms the phospholipid molecule into a new soap-like molecule that attracts and binds fat and thereby ruptures cell membranes. *Cardiotoxins* are specifically toxic to the heart. They bind to particular sites on the surface of muscle cells and effectively prevent muscle contraction. These toxins may cause the heart to beat irregularly or stop beating completely. *Hemotoxins* cause hemolysis, the destruction of red blood cells (erythrocytes). Another mechanism induces blood coagulation. A common family of hemotoxins are snake venom metalloproteinases. There are also oligopeptides, which inhibit angiotensin converting enzyme (ACE) and thereby lower blood pressure.

From an evolutionary perspective it is known that venoms developed approximately 170 million years ago, and then diversified into the huge venom diversity seen today.[168] What has been developed and refined by nature over so many years can now be used by humans to develop drugs.

Recent and very promising new developments of venom-based peptide therapeutics include FDA approved antidiabetic and analgesic drugs, as well as promising monomeric insulins, treatments of autoimmune diseases and (spider) toxins for use as eco-friendly insecticides. Table 6B.1 shows a list of approved drugs derived from animal venoms.[169]

A review of promising animal venom-based cancer treatments, published recently by Rui Ma et al.,[170] and Bin Li et al.[171] reported how *scorpion venom* could be used to fight cancer. Rui Ma concluded that the initial interaction of venom peptides with the target molecule is the first and foremost step to induce anti-cancer activity. Followed by initial interactions, peptides tend to exhibit their effects mostly by membrane interactions, or by intracellular peptide-protein interaction and peptide-DNA interaction. Currently, venom-based drugs such as chlorotoxin and integrin $\alpha v \beta 3$ drugs are used mainly in brain tumor and cancers with overexpressed $\alpha v \beta 3$.

Table 6B.1 Approved drugs from animal venoms (adapted from G. F. King's review)

Protein	Drug name	Animal source	Molecular target / delivery	Indication	FDA appr.	Company
Captopril	Capoten	Pit viper	ACE	Hypertension	1981	B-M Squibb
Eptifibatide	Integrilin	Pygmy rattlesnake	Alphaiiib Beta3 integrin receptor	Acute coronary syndromes	1998	Merck
Tirofiban	Aggrastat	Saw-scaled viper	Alphaiiib Beta3 integrin receptor	Acute coronary syndromes	1999	Iroko & Merck
Bivalirudin	Angiomax	Medicinal leech	Thrombin	Coagulation during surgery	2000	The Medicines Co.
Ziconitide	Prialt	Cone snail	Cav2.2 channel	Chronic pain	2004	Azur & Eisai
Exenatide	Byetta	Gila monster lizard	GLP-1 receptor	Type 2 diabetes	2005	Amylin & Lilly
Batroxobin	Baquting	Lancehead snake	Fibrinogen	Perioperative bleeding, thrombotic disorders	Outside USA, China	Nuokang Biopharma

There is also the direct way of acquiring immunity against venoms by "training" the immune system, following the saying "what does not kill you makes you stronger." Among animals, the hedgehog, the mongoose, the honey badger, the secretary bird, and a few other birds that feed on snakes appear to have acquired immunity to a dose of snake venom. Among humans, there is the interesting story of Bill Haast, the late owner and director of the Miami Serpentarium. He injected himself with snake venom during most of his adult life, in an effort to build up an immunity to cobras and other venomous snakes. What Haast did is known as *mithridatism*. Haast lived to age 100[172] and survived a reported 172 snake bites. It seems that immunity to snake venoms also strengthened his immune defense against other diseases. He frequently donated his blood to snake-bite victims and saved at least twenty such individuals.

Let's now turn to specific animal–human health stories, starting with the pig.

The Domestic Pig (Sus domesticus)

"Little pig, little pig, let me come in."
"No, no, by the hair on my chiny chin."
"Then I'll huff, and I'll puff, and I'll blow your house in."[173]

Pigs include the domestic pig (Fig. 6B.1) and its ancestor, the common Eurasian wild boar (*Sus scrofa*). Pigs are native to the Eurasian and African continents. They are highly social and intelligent animals. There are more than 1 billion domestic pigs alive at any time. China has the largest pig population, above 400 million, and Denmark probably has the highest number of pigs per capita, with around 25 million compared to ~5 million Danes. Pigs are omnivores and can consume a wide range of food. Biologically, pigs are very similar to humans, and thus are frequently used for human medical research.

Pigs were domesticated by humans about seven thousand years ago. By comparing the genomes of domestic pigs and wild boars, the evolution of pig genetic groups during domestication can be identified. In the genome of the domestic pig, there has been a significant expansion of the immunity- and olfactory-related genes: 1301 olfactory receptor genes and 343 preferred olfactory receptor genes have been identified in the pig's genome. The preliminary explanation of the genetic mechanism of olfactory sensitivity in pigs reflects the pig's high dependence on the sense of smell.

Figure 6B.1 The domestic pig, a beloved human companion and medical research model (Photo by Michael H./Drawing: Wikipedia).

Domestic pigs are raised commercially as livestock. Their meat is known as pork, the skin is used for leather, and their bristly hairs are used to make brushes. Because of their excellent sense of smell, they are used to find truffles in Southern European countries. Some religions (Judaism, Islam) consider the consumption of pork sinful. On the other hand, most Christians do not consider pigs as "unclean." In many European countries, the slaughtering of a pig is celebrated as a feast. In Germany, pigs are known as a symbol for good luck. Marzipan pigs are handed out as gifts on New Year's Eve. English bedtime stories about pigs are read to children.

The physiological structure, behavior, and nutritional needs of domestic pigs are similar to those of humans and can be used to study human diseases. Domestic pigs are highly susceptible to influenza, which is a threat to the food supply (e.g., highly contagious African swine fever outbreak 2018 in China), but also of great benefit to the development of human influenza vaccines.

An interesting observation was made by H. C. Gerstein and L. Waltman,[174] who attempted to explain why domesticated pigs do not develop type 2 diabetes. There seems to be general agreement that the current global diabetes epidemic is occurring because our hunter-gatherer ancestors evolved in an environment with uncertain food availability and high demands for physical activity, whereas we live in a contemporary world with an abundance of food and decreased requirement for physical activity.

Domesticated mammals such as pigs are the result of generations of selective breeding that targeted a phenotype best suited for a particular task. Thus, for thousands of years, domesticated pigs and cows were selectively bred for their ability to efficiently accumulate and store energy for later consumption by humans, whereas dogs and rodent-catching cats were selected to maximize physical work and minimize consumption of food supplied by humans. Pigs and cows should therefore be protected against the toxic effects of a "diabetogenic" environment characterized by inactivity and energy abundance. On the other hand, dogs and cats would have no such protection. Indeed, hyperglycemia is either unreported or extremely rare in pigs and cows and other domesticated mammals that get minimal exercise and are fattened quickly to be slaughtered, whereas it is relatively common in dogs and cats. Observations from pig studies suggest that susceptibility and resistance to diabetes may be related to quite recent environmental effects. About 500 years ago, a colony of domesticated pigs was released on Ossabaw Island, off the coast of Savannah, Georgia, USA. Since then, they have lived in an environment characterized by an uncertain food supply and high physical demands. Despite the shared ancestry, when these pigs are captured and raised together with their "food-producing cousins" in a high-calorie, low-activity environment, they experience obesity and hyperglycemia, whereas the other pigs do not.[175]

This would also explain why people of European descent are more resistant to diabetes than most other ethnic groups in a Westernized "diabetogenic" environment. For at least 300 years Europeans have enjoyed a relatively stable food supply and availability of labor-saving devices. Other populations that are newly exposed to such an environment may have little resistance to its diabetogenic effects and will experience rising diabetes rates.

The physiological similarity between pigs and humans has been beneficial during the early days of diabetes treatments with insulin: Since Banting's breakthrough research, awarded with the 1923 Nobel Prize,[176] insulin has been used to help diabetes patients. The early leaders in insulin production, Novo Nordisk of Denmark and Eli Lilly of Indiana, all started out in the vicinity of slaughterhouses, where they could extract insulin from the pancreas of pigs and cows. In the 1980s things changed, nowadays insulin is produced by means of genetic engineering (see Chapter 7).

Recently, medical research based on pigs has intensified. Increasingly, genetically modified pigs[177] are used. The Danish team led by Lars Bolund has been pointing out that pigs have been used to develop some of the most important large animal models for biomedical research. Advances in pig genome research, genetic modification of primary pig cells and pig cloning by nuclear transfer, have facilitated the generation of genetically modified pigs. They are already used for organ and tissue transplantation and for medical research into various human health issues including neurodegenerative diseases, cardiovascular diseases, eye diseases, bone diseases, cancers, epidermal skin diseases, cystic fibrosis, diabetes mellitus, and other metabolic diseases.

Regenerative medicine is another research field where pigs hold a lot of promise.

In 2014, Kitamura et al.[178] used pig bladders as scaffolds to regenerate the larynx of dogs. A partial laryngectomy was performed in five beagle dogs, and immediately after, the pig bladder scaffold was fitted to the surgical defect and sutured. Functional data were obtained 6 months after intervention and were within normal range in three of the five animals studied. Histological analysis revealed regeneration of cartilage, vocal fold mucosa, and muscle. More importantly, the scaffold supported regeneration of the different tissues in their original positions separately—namely, the defect area was populated by cartilage, muscle, and epithelium that connected to other cartilage, muscle, and epithelial tissue types around the injury site. Kitamura et al. believe that the scaffold recruited different cell types from the vicinity of the injury to regenerate the dogs' larynx tissue.

Later in 2014 it was shown by a team at University of Pittsburg that similar pig bladder scaffolds could be used to regenerate leg muscles.[179] Although muscle has the ability to regenerate naturally, it cannot refill massive defects, such as those seen in volumetric muscle loss (VML). In response, the team implanted a biomaterial scaffold at the site of VML, encouraging local muscle regeneration and improving function in both mice and humans. The biomaterial used in this study was made up of pig bladder tissue that had been stripped of cells, leaving behind only the protein scaffold called the extracellular matrix (ECM). There was new skeletal muscle formation including striated (striped) tissue organization. The new muscle was also was innervated, which is necessary for function. The preclinical work was then translated into a clinical study of five patients with VML and outcomes were similar to the mice. Six months after ECM implantation at the site of muscle loss, all patients showed signs of new muscle and blood vessels. Three of the five patients showed 20% or greater improvement in limb strength during physical therapy.

This proves that pigs can be used as "translational research models," which link basic science to clinical applications in order to establish novel therapeutics. Scientists have learned that "what works in the pig has a high possibility of working in sapiens."

For instance, a pig's heart is about the same size and shape as a human heart. Pigs develop atherosclerosis—artery plaque buildup—in the same way that humans do, and they react similarly to myocardial infarction, the classic heart attack. Because of these similarities, scientists have long used pigs to test interventional catheter devices and methods of cardiovascular surgery. Also, tissues derived from pig hearts have been used to replace defective heart valves in humans, lasting upwards of 15 years in the human body.[180]

Beyond their closely related hearts and blood vessels, another characteristic humans and pigs share is their diet. Both eat meat and plants to survive. The pig's physiology of digestion and the metabolic processes in the liver are also similar to humans. Pigs are therefore used in a lot of dietary type of studies, as well as oral absorption studies of drugs.

Similarly, pig kidneys are comparable in size and function to human kidneys, lending themselves to renal research.

In dermatology, pigs have been one of the standard plastic surgery models because their skin wounds heal similarly to human skin.

Let's return to diabetes, this time type 1 or juvenile diabetes where the beta cells in the pancreas no longer produce insulin. As the insulin-producing cells in a pig's pancreas are similar to humans', a significant amount of research on diabetes has been aimed recently at isolating those cells and harnessing them for future treatments.

Because of the similarities in organ systems and the growing problem of human donor organ shortages, pigs have also been targeted as potential organ donors for humans. Though primates such as baboons and chimpanzees are more closely related to humans, pigs are much more readily available and may be ethically more acceptable.

There are ongoing NIH funded studies of pig-to-primate organ transplantation.[181] In 2016, a team led by Dr. Mohiuddin,[182] worked with a group of pigs that were genetically modified in several ways. First, they lacked a key molecule known to provoke organ rejection. The pigs were also engineered to produce human proteins that suppress blood clotting and activation of other common causes of xenograft failure. The team transplanted 5 pig hearts into baboons. Instead of replacing the baboon hearts, they implanted them in the baboon's abdomens so that the baboon's own hearts continued to pump blood. The scientists then developed an immune-suppressing regimen using an immunomodulatory drug (mycophenolate mofetil) and antibodies against key immune system components (CD40 and CD20). With this regimen, the pig hearts survived in the baboons for a mean of 433 days, the longest lasting over 2 and a half years.

Due to this high risk of transplant rejection, there may be another option, namely to grow organs based on human stem cells inside pig organs.

Research along those lines is going on at the Texas Heart Institute's Regenerative Medicine Research Lab in Houston, Texas.[183] In 1969, Denton Cooley, founder of the institute, was the first to implant an artificial heart into a human. Today, Doris Taylor and her 25-person team are working on the idea to implant a very special pig-based heart into tomorrow's patients. It has been scrubbed clean of all cells, leaving only collagen, fibronectin and laminin, to provide a protein scaffold on which to build a new human heart. As a first step, Taylor's multidisciplinary team members decellularize seven or eight hearts (from rats) a week, then inject the DNA-free scaffolds with stem cells. Because muscular heart cells do not divide, they cannot regenerate on their own like, say, the bladder, an organ which has been regenerated and implanted in humans. In the case of the heart, stem cells (as opposed to heart cells) adhere to the surface of the scaffold, growing into living, functioning organs inside machines known as bioreactors, which replicate the warm, oxygen-rich environment of a heart inside a mammal's body. Dr. Taylor estimates that it will take 10 to 15 years before a functioning heart, based on a pig heart scaffold, will be implanted into an adult human. As is often the case when doing leading edge research, beneficial side effects in form of new insights into regenerative medicine will be expected.

Although those stories only indicate first steps of what perhaps may become future medical practice, they prove the importance of the pig not only as a human food source, but as a potentially life-saving medical resource.

As to today's medical practice, there is already one area, namely bone grafts, where the bones of pigs are utilized in a clinically approved way. Bone grafts can be done in different ways, namely

- *Autografts*: graft material harvested from the patient's own body. However, failure rates are high due to bone loss: the transplanted bone may not survive in the new body location.
- *Allografts*: bone graft material taken from another human, from a tissue bank. However, people donating their bone may carry a disease.
- *Xenograft*: bone graft material comes from another species such as bovine (cow) or porcine (pig) bone. The inorganic part of the bone is used as a scaffold, leading to minimal risk for complications.
- *Alloplast*: bone graft material that is synthetic, typically based on resins, hydroxyapatite, calcium phosphate, and other minerals that allow patient's bone to regenerate.

Xenografts are used as dental implants and in other surgical applications, and porcine bones are considered to be very safe.

An important initiative associated with pigs is pursued by eGenesis, a startup co-founded by George Church and Luhan Yang (Harvard University). In August 2018, eGenesis published the successful inactivation of Porcine Endogenous Retrovirus (PERV) in pigs.[184] By producing the first PERV-free piglets, the risk of cross-species viral transmission can be avoided. This scientific break-through represents an important milestone for xenotransplantation, the use of animal organs for human transplants. Given the huge human organ transplantation bottleneck, a successful translation into clinical practice would meet an unmet medical need. More about the genetic editing part in Chapter 7.

Before closing this chapter, let's hear what role the pig is playing in Chinese culture.

An important biological characteristic of domestic pigs is their strong fertility. In China, pigs are associated with both fertility and virility. Chinese couples who are trying to have children may even display pictures of pigs prominently in their bedrooms.

In the Chinese zodiac, the pig represents fortune, honesty, and happiness. How appropriate for this honest, happy animal that is smart, lovable, and forgiving.

Here are a few additional facts about pigs.[185]

Pigs are smart and have good memory: Pigs are smarter than dogs, some primates, and even (up to) three-year-old children. Their level of cognitive ability allows pigs to recognize their own names, to dream, and to follow commands. Pigs can use mirrors to locate food that is not directly visible, a task only a few other animals such as dolphins, elephants, and chimps can accomplish. Their ability to understand how other pigs think allows them to use deceit and purposely mislead other pigs to the wrong trail so they can keep more food for themselves. Pigs also have excellent memories. Studies have shown that pigs can remember where food is stored and places where they have found food before. They can find their way home from great distances. Pigs can recognize and remember humans and up to 30 other pigs.

Pigs are social: They form close bonds with people and other animals. They love contact and enjoy getting massages. They greet each other by rubbing noses and enjoy hanging out together, sunbathing, playing, and listening to music. Pigs love to sleep together cuddled up nose-to-nose.

Pigs talk a lot: They communicate constantly with each other and have a vocabulary of over 20 distinct oinks, grunts, snorts, and squeaks that have specific meanings. They convey their intentions, how they are feeling, warnings, greetings, and when it's time for dinner. For instance, mother sow has a special call to let her piglets know that it is time to suckle. Newborn piglets learn to recognize their mothers' voices. Mother pigs also talk (or perhaps "sing"?) to their piglets while nursing.

Pigs have feelings: They have individual personalities with a wide range of traits and emotions. Some pigs are more serious, some are more daring, and some are more introvert. Like humans, they may even suffer from depression. They can feel happiness, sadness, grief, and pain. Pigs are highly sensitive animals and can become anxious and depressed when confined to cramped spaces and mistreated. Pigs are peaceful animals but may show aggression when their young are threatened. Pigs are mostly happy animals who like to have fun.

The (not so happy) reality: Most of the pigs raised in factory farms are confined and will never get to play, be clean, cuddle with each other, or be picky eaters. The females will be forcibly impregnated, kept in gestation crates, have their babies taken away from them, and eventually be slaughtered. The piglets will never fully experience the love of their mothers. They will be kept in crowded, often dirty, conditions with no ability to move. Even in the best conditions of the friendliest farm, pigs raised for food will be killed and not allowed to live to their natural lifespan with their families in peace.

Pigs are wonderful animals and deserve to live a longer and happier life. However, humans are not as sensitive and compassionate as pigs.

A quote[186] attributed to Winston Churchill: "I am fond of pigs. Dogs look up to us. Cats look down on us. Pigs treat us as equals."

In conclusion, there are many reasons to love pigs—not only because of the delicious Spanish Iberico ham, the Italian prosciutto, their long-time human connection, their long history of helping diabetes patients, and the future promise of organ transplantation.

And one more thing: since pigs are so similar to humans, have good memory but don't seem to get Alzheimer's disease, is there something else we could learn from them? Or does it have something to do with the fact that humans don't let pigs get old enough?

The Elephant (Loxodonta africana and Elephas maximus)

According to recent data collected by the Convention on International Trade in Endangered Species (CITES[187]) of Wild Fauna and Flora, there are 415,000 elephants in Africa. Major threats to the elephant population are caused by poaching. Trophy hunting remains controversial but is not considered to be the biggest threat to wildlife, at least not in regions that have well-functioning governments.

Elephants are the world's largest land mammals and, aside from the great apes, the most intelligent animal. Elephants are smart, social, and incredibly strong creatures. An elephant's trunk is so powerful and precise that it can carry calves and also be used for more delicate acts like picking flowers! Male African bush elephants (see Fig. 6B.2) average a height of 3.20 m at the shoulder and weigh about 6000 kg. Their Asian counterparts (Fig. 6B.2) are smaller, with males being 2.75 m tall at the shoulder and weighing 4000 kg on average. Female African elephants are typically 23% smaller than males, whereas female Asian elephants are only around 15% smaller than males. While huge among terrestrial life forms, the large weight of elephants pales in comparison to the weight of blue whales that can exceed 130,000 kg.

Figure 6B.2 "Vulnerable" African (large ears) and "endangered" Asian (smaller ears) elephant (source: Wikipedia).

Elephants eat leaves, twigs, fruit, bark, grass, and roots. They can consume as much as 150 kg of food and 40 L of water in a day. They have a huge impact on their environments. For instance, their habit of uprooting trees and undergrowth can transform savannah into grasslands. When they dig for water during drought, they create waterholes that can be used by other animals.

Elephants average only 3–4 hours of sleep per day. Herds of elephants typically move 10–20 km a day, but can cover more than 100 km if needed. Elephants are also known to go on seasonal migrations in search of food, water, and mates.

The social structure of elephants differs significantly between females and males. Female elephants can spend their entire lives in matrilineal family groups, sometimes counting more than ten members. Such family groups may include three pairs of mothers with offspring, and are frequently led by the eldest female, the "matriarch." She remains leader of the group until death or if she no longer has enough energy to play her role. When her reign as matriarch is over, her eldest daughter steps up, even when the matriarch's younger sister may still be around. During the dry season, elephant families may cluster together and form another level of social organization known as the "clan." Groups within these clans do not form strong bonds, but they defend their dry-season ranges against other clans. A clan may hold up to 90 elephants, i.e., nine family groups.

The social life of the adult male is quite different. As he matures, a male spends more time at the periphery of his group and associates with outside males or even other families. By the age of 15, young males may spend over 80% of their time away from their families. When male elephants leave their family permanently, they either live alone or with a few other males. Larger bull groups consisting of over 10 members occur only among African bush elephants. These elephants can be quite sociable when not competing for dominance among mates. They may also form long-term relationships. The dominance hierarchy among mates depends on their age, size, and sexual condition. When in groups, males follow the lead of the dominant bull. Young bulls typically seek the company and leadership of older, more experienced males.

Gestation of elephant babies takes two years. Calves are totally dependent on their mothers for food (milk) during their first 3 months. They then start mixing in other food but continue to suckle at the same rate as before until their sixth month. By nine months, mouth, trunk, and foot coordination is fully developed. After its first birthday, the calf still needs its mother for nutrition and protection from predators for at least another year. Suckling then may last another year. Female calves are sexually mature by the age of nine years, while males need another 5–6 years to become mature. Elephants have a long life expectancy, reaching up to 80 years of age.

Anatomically, perhaps the most interesting feature of elephants is the "trunk." The trunk is an elongated fusion of the elephant's nose and upper lip. It is extremely versatile, containing up to 150,000 separate muscle fascicles (bundles of muscle fibers surrounded by connective tissue), with no bone and little fat. Elephant trunks have multiple functions, including breathing, smelling (olfaction), touching, grasping, and sound production. The elephant's sense of smell is highly developed exceeding the well-known capabilities of dogs. The trunk's ability to make powerful twisting and coiling movements permits the collection of food, wrestling with other elephants, and lifting weights of up to 350 kg. In addition, the trunk can be used for delicate tasks, such as wiping an eye or cracking a peanut shell without breaking the seed.[188] Helped by its trunk, an elephant can reach items at heights of up to 7 m as well as dig for water covered by mud or sand. The elephant's nose can also be used to absorb water. When the elephant is thirsty, it puts its nose into the river to suck water, functioning like a small pump. To avoid the mix of air and water, there is a piece of special cartilage above the esophagus behind the elephant's nasal cavity, acting like a valve. As the trunk absorbs water, the muscles in the throat area contract and the "valve" closes, allowing the water to enter the esophagus without entering the trachea. After drinking water, the residual water in the nose is ejected. At this time, the "valve" is automatically opened and the breathing is performed normally. Elephant life cannot be separated from water: on hot days, they use the nose to suck enough water, and then spray the body like a shower bath. In addition, they can apply mud or sand to the body to prevent mosquito bites and protect the skin.

The tusk is used as a weapon in battles with rivals and as a courtship aid: the larger his tusks, the more attractive a male elephant may appear to a female.

Another interesting aspect of elephant life is associated with thermal regulation and protection against intense solar radiation, in particular for the African bush elephant. In a recent paper in *Nature*[189] by Antonio F. Martins et al., medical imaging and computer simulations were used to show that the intricate network of crevices on the skin surface of elephants is caused by fractures (due to bending stress) of the animal's brittle outermost skin layer. The resulting micrometer-wide channels increase water retention and mud adherence, thereby protecting the elephant against parasites and solar radiation. The authors further suggest that there is a parallel between physiological characteristics of elephant skin and the skin of humans affected by *ichtyosis vulgaris*, an inherited skin disorder in which dead skin cells accumulate in thick, dry scales on the patient's skin surface. The condition is sometimes called fish scale disease or fish skin disease and often goes undiagnosed because it is mistaken for extremely dry skin.

Elephants produce low-frequency sounds at high amplitudes, producing acoustical waves or "seismic signals" that can travel along the surface of the earth. They use those seismic signals for communication. When transmitting or receiving them, elephants appear to rely on their leg and shoulder bones to transmit the signals to the middle ear. They lean forward and put more weight on their larger front feet, thereby utilizing the cushion pads of the feet to pick up seismic communication. Elephants may be able to detect a thunderstorm from 280 km away, and will head towards it, looking for water. In 2004, elephants appeared to head for higher ground *before* the Asian tsunami struck.

Elephants are also famous for their many vocalizations. This includes chirps, trumpets and high-pitched squeaks that can travel far to other groups. Those vocalizations are used for both daily communication and to signal willingness to breed.

Elephants have always inspired human respect and admiration. In many cultures, elephants represent strength, power, wisdom, longevity, stamina, leadership, sociability, and loyalty. They have been used as war animals, for work and transportation, but also as zoo attractions and as circus animals. Similar to apes and dolphins, elephants recognize themselves in mirrors. They are known to use tools and have long been admired for their good memory: their ability to form cognitive maps allows them to remember extended spaces over long periods of time, including an ability to keep track of the current location of their family members. Their brilliant memory also allows them to remember good waterhole and feeding locations.

Elephants appear to understand what other elephants are feeling. Experiments show that when one elephant is unhappy, others share their feelings, something known as "emotional contagion." In these situations, they will go over to their "friends" and comfort them, often by putting their trunk into the other's mouth, something that elephants find reassuring. Elephants will also assist other injured elephants, and even appear to mourn their dead.

Elephants and humans have some interesting things in common. Both have a long life span, strong family ties, and understand death. Also, elephants are born with their brains weighing about 35% of their adult brain weight and humans are born with their brain weighing about 25% of their adult brain weight.[190] In contrast, most mammals are born with brains weighing about 90% of their adult brain weight. So, both elephants and humans do most of their brain development after birth. As the largest land mammal, the elephant also has the longest pregnancy at 22 months! Newborn calves weigh around 90 kg and are less than 1 m tall.

An international group of researchers sequenced the complete genome of both extinct and living elephantid species[191] and saw evidence of genetic admixture among multiple species. In addition, the team found that the African forest and savannah elephants exist as separate species. In the study, the team's genome-wide analysis demonstrated that about 500,000 years of isolation occurred between their respective ancestors.

Through sequencing, the researchers found that the straight-tusked elephants descended from a mixture of three ancestral lineages. Most of the species' ancestry derives from an African elephant ancestor, while its remaining ancestry consists of a large contribution from a lineage related to mammoths, and the other related to still existing ("extant") forest elephants. The analysis of whole elephant genomes highlights that multiple major interbreeding events occurred between different ancient species. The team believes future studies should explore whether admixture was not only an important phenomenon in the demographic history of elephantids, but if it also played a biologically important role in their evolution.[192]

In the context of our theme for this book, it may be worth exploring the longevity of elephants in more detail. Although there are exceptions to this rule, there is a lot of statistical data indicating that bigger animals live longer.[193] Perhaps even more interesting is the related fact that elephants do not get cancer as easily as humans.

In a research paper published in 2016,[194] scientists representing the Universities of Chicago, Weill Cornell in New York, and Nottingham, UK, focused on "TP53 copy number expansion" as a key factor in protecting large animals like elephants from enhanced DNA damage and associated cancer risk.

TP53, the gene that encodes protein p53, is also known as "tumor suppressor gene" and sometimes referred to as the "guardian of the genome." Human cancer is strongly associated (more than 50%) with mutations of the *TP53* gene. When functioning properly, the

TP53 gene encodes p53 proteins that bind to DNA and regulate gene expression to prevent mutations of the genome. The p53 protein senses when DNA is damaged or a cell is under stress. Then it either slows the cell's growth while the damage is repaired, or it triggers cell death. Inactivation of the TP53 gene therefore causes suppression of apoptosis (a process of programmed cell death that occurs in multicellular organisms), increased (cancer) cell proliferation, and genomic instability. The crucial importance of TP53 for cancer risk is illustrated by the fact that humans with only one allele of TP53 run a 90% lifetime risk for cancer, for multiple tumors, and early-childhood cancers. Understanding the cellular mechanisms of cancer suppression in animals could therefore benefit humans at high risk of cancer, and even the healthy, aging population.

While humans only have one copy of TP53, elephants have several versions and get cancer at a much lower rate than humans. We conclude that cancer risk is decreased in living organisms that carry several extra copies of TP53, explaining Oxford University statistician Richard Peto's[195] paradox (stated in the late 1970s): the incidence of cancer (at the species level) does not appear to correlate with the number of cells in an organism. Every time a cell divides, it may gain a mutation that causes cancerous growth. The bigger an animal is, the more cells it has, and the longer an animal lives, the more times its cells divide. However, the incidence of cancer in humans is much higher than the incidence of cancer in elephants and whales, despite the fact that those huge animals have many more cells than humans. If the probability of carcinogenesis were constant across cells, one would expect large animals to have a higher incidence of cancer than humans. Therefore, "massive animals [such as elephants] may hold secrets of cancer suppression."[196]

Due to the high potential impact of research into elephant's ability to suppress the risk of cancer, several research groups are currently engaged in attempts to translate our increased understanding of TP53 "copy number expansion" into clinical studies that ultimately may benefit human health. For instance, a study led by Joshua Schiffman's research team at the Huntsman Cancer Institute at the University of Utah and published in the Journal of the American Medical Association (JAMA),[197] found that, compared to humans, elephants have up to 20 times the number of tumor suppressor genes TP53. Dr. Schiffman and his colleagues reviewed zoo records on the deaths of 644 elephants and found that less than 5 percent had died of cancer. By contrast, 11 percent to 25 percent of humans die of cancer—despite the fact that elephants can weigh a hundred times as much as humans do.

Instead of trying to repair damaged DNA within cells, some elephants' p53 proteins instead are killing a substantial number of cancer cells, thereby protecting them from the disease. As recently reported in *Science*,[198] elephants also carry a gene that induces apoptosis whenever p53 detects a DNA copy irregularity. Details are published in *Cell Reports*,[199] speculating that elephants and their extinct relatives (proboscideans) may have resolved Peto's paradox in part through re-functionalizing a leukemia inhibitory factor pseudogene (LIF6) with pro-apoptotic functions. LIF6 is transcriptionally upregulated by TP53 in response to DNA damage and translocates to the mitochondria where it induces apoptosis. The analysis of living and extinct LIF6 genes indicates that its TP53 response element evolved coincident with the evolution of large body sizes![200]

Based on these findings, researchers at the Technion-Israel Institute of Technology in Haifa are conducting a follow-up study in an effort to translate the discovery into innovative cancer treatments.[201]

In children's books, elephants are frequently cast as models of exemplary behavior. They are strong but kind, they don't kill or prey on other animals, and elephant mothers protect their children, and may even display affection and sorrow similar to humans. In addition, they may be able to teach us how to prevent and fight cancer!

And here is another speculation: how about the excellent long-term memory exhibited by elephants. There does not seem to be any occurrence of dementia or Alzheimer's disease. Another opportunity for humans to discover why—and to learn?

The Naked Mole Rat (Heterocephalus glaber)

The naked mole rat (Fig. 6B.3) is a burrowing rodent native to parts of East Africa. Along with the Damaraland mole rat, it is the only known eusocial mammal, the highest classification of sociality, otherwise only found in insects. It has a highly unusual set of physical traits that allow it to thrive in a harsh underground environment. Typical individuals are 8–10 cm long and weigh 30–35 g. Queens are larger and may weigh 50–80 g. Their legs are thin and short but highly functional at moving underground. Their large, protruding teeth are used to dig. Their lips are sealed just behind the teeth, preventing soil from filling their mouths.

Figure 6B.3 Naked mole rat, remarkable for its longevity and resistance to cancer and hypoxia (source: Wikipedia).

The naked mole rat gets its name from its hairless body and lives in a dark, low-oxygen, high-CO_2 environment for life. Actually, mole rats are not completely naked: they have about 100 hairs on their bodies that can be used as whiskers to help perceive. The naked mole rat has almost completely lost its vision (as opposed to the Spalax mole rat, which is truly blind). It only uses the tentacles on both sides of the body to identify the direction. Although it is a mammal, it cannot produce heat. Like cold-blooded animals, it can only regulate body temperature through heat exchange with the environment. Its skin cannot feel pain stimuli and the mole-rat has gained fame by showing high immunity to cancer.

The naked mole rats are sometimes accused of being ugly—certainly a matter of taste—but it is definitely hard to embrace their habit of eating the feces of the naked mole rat queen. This is indeed a bit unusual. Although rabbits, rodents (mouse, rat, hamster, nude mole, chinchillas, and guinea pigs), dogs, mountain beavers, elephants, hippos, and monkeys such gorillas, orangutans, and macaques also have a hobby of eating feces, they mostly do it to better absorb the nutrients that were not fully digested when they first ate. In addition, there could also be the reason to avoid predators by "destroying the traces." However, mole rats are doing it for a different reason, as discussed further below.

Though naked mole rats can live for more than 30 years (up to 20 times as long as comparable-sized rodents), they almost never emerge from below ground.[202]

The naked mole rat was first documented by Eduard Rüppell, a German naturalist, in the 19th century, but he assumed from its unprepossessing appearance, the sagging nude flesh, the teeth growing straight through the skin, that it was a diseased or mutated individual of another species. It was not until the 1950s that the strange habits of this animal began to be known.

Here are some interesting facts[203] about these creatures and their amazing evolutionary traits.

- Although mole rats are members of the rodent family, they are not rats. Their closest relatives are guinea pigs, porcupines and chinchillas. They live in subterranean colonies in East Africa.

- Mole rats are eusocial, like bees and ants: There is a mole rat queen and only two or three males responsible for reproduction. The rest of the mole rats have specific roles to play. They are digging tunnels, finding food, or fending off predators like snakes. A mole rate colony includes up to 300 naked mole rats.

- The mole rat queen is a ferocious fighter. Queens get to their position by fighting off other females and when a queen dies or is considered a bit weak. During the battle to find her successor, the competing mole rats will use their long, protruding incisors to stab and kill opponents. The winner will then grow bigger by enlarging gaps in her vertebrates so she can bear offspring. The queen will also secrete pheromones in her urine to render the competition infertile. To remain the queen, a female mole rat is required to reproduce several times a year and will literally be trampling on her competition and her subjects to show dominance. There is more about the queen's total domination: In other female mammals, child-rearing instincts are triggered by hormones during pregnancy. However, these hormone-producing reproductive organs never develop in the subordinate females in naked mole rat society. So what is the substitute? According to a Japanese study published August 2018[204,205] in the *Proceedings of the National Academy of Sciences* (PNAS), the answer could be related to the eating of the queen's poop: when comparing three groups of female subordinates, one eating the pregnant queen's dung, the second dining on poop from a nonpregnant queen, and a third eating feces from a nonpregnant queen that had been enhanced with an estrogen hormone, it was found that groups 1 and 3 exhibited similar behavior: they turned their heads to listen to mole rat pup cries, and had higher levels of estrogen in their poop and urine, than did the other mole rats. Hence naked mole rat queens pass along maternal instincts to their subordinates through fecal meals—a strategy probably unique in the animal kingdom.

- Naked mole rats live entirely underground. For them it's not a problem, evolution has made sure they can cope with the conditions. The tiny eyes and almost completely sealed up ears are protected from the soil, and their teeth grow outside the mole rat's mouth to prevent it from eating soil as it uses them to dig new tunnels. They have no sweat glands, fur or fat, so they do not overheat in close quarters underground. They are able huddle closely together to keep warm. They have tiny lungs but are using oxygen incredibly efficiently because of their ability to bind more oxygen in the blood. They have a very low metabolic rate, conserving energy to such an extent that they can live up to 32 years.[206] No other rodent lives as long.

- They deal extremely well with hypoxic conditions. An international team led by Thomas Park, University of Illinois at Chicago, has found how the mole rat borrows a technique from the plant world to survive without taking on new oxygen supplies.[207] By changing from an oxygen-reliant glucose-based metabolic system to one that uses fructose, the mole rats can live while being deprived of oxygen. The naked mole rat survives for at least 5 hours in air that contains only 5% oxygen; it does not show any significant signs of distress and continues normal activity. Even in zero-oxygen conditions it can survive up to 18 minutes.

- Surprisingly, mole rats have excellent hygiene. They have burrows within colonies that are dedicated to nesting, eating, and even farming. While digging tunnels they look for roots and pull them into burrows. If they find a large tuber,[d] they eat it slowly, covering it up again so it continues to live. They also have a toilet burrow, which every naked mole rat uses. Once it's full, they close it off and dig a new one.
- They can run backwards as fast as forwards. When they run, their incisors can move independently of each other like chopsticks.
- Biologists recently discovered that an acid burn did not seem to cause pain for mole rats. When a research team used acid and a chemical that makes chilis hot, to stimulate the rodent, they found that the nerve signals went to the "touch region" rather than the pain region. New research published in *Cell Reports* in 2016[208] shows that evolutionary mutations are responsible. Normally, when sensory neurons are exposed to high temperatures and inflammatory attacks, nerve growth factor (NGF) molecules bind to a receptor called TrkA. This triggers the TRPV1 channel to open, and once the channel is opened, the neuron tells the brain that pain has occurred. The naked mole rats differ from other animals in this process: their TrkA receptors show little change. A comparison of the amino acid sequences of the TrkA receptor in nude mice and normal rats revealed that there were three amino acid differences, resulting in down-regulation of pain perception!
- Naked mole rats do not get cancer. Up until 2016, not one case of cancer among naked mole rats had ever been reported. The first ever discovered malignancies in two naked mole rats were reported in two individuals that were captive-born at zoos, and hence lived in an environment with 21% atmospheric oxygen, much higher than their natural 5–9%.

A possible mechanism involved in cancer protection may be p16, a tumor suppressor protein, that in humans is encoded by the

---

[d]A thickened underground part of a stem, e.g., in the potato, bearing buds from which new plants arise.

CDKN2A gene and plays an important role in cell cycle regulation. It prevents cell division once individual cells come into contact (known as "contact inhibition"). The cells of most mammals, including naked mole-rats, undergo contact inhibition via the gene p27, which prevents cellular reproduction at high cell densities. The combination of p16 and p27 in naked mole rat cells is a double barrier to uncontrolled cell proliferation,[209] one of the hallmarks of cancer and is known to be implicated in the prevention of several cancers, notably melanoma.

The same biology team from the University of Rochester, New York,[210] discovered that the matrix that supports tissues of a naked mole rat is full of a substance called hyaluronic acid, which both acts as a lubricant and also happens to prevent cancer from growing. Cancer starts when mutated cells grow out of control. Hyaluronic acid, which has been described as "gooey sugar," keeps these tumor cells from replicating. The Rochester group found that if it prevented the mole rat tissue from generating the lubricant, it did in fact grow tumors. A few months later, the same University of Rochester research team also announced that naked mole rats have ribosomes that produce extremely error-free proteins.[211]

It is speculated that naked mole rats are combining their cancer preventing weapons to more or less deactivate tumor cells before they can get out of control. Studying naked mole-rats could therefore provide valuable insights for the prevention of cancer in humans.

From what we have learned about mole rats, it is not surprising that they have attracted a lot of attention, mostly by research teams interested in cancer and aging.

Calico (acronym for "California Life Company"), a biotech company founded in 2013 by Bill Maris and backed by Google with the goal of combating aging and associated diseases, has shown a strong interest in naked mole rats. In 2018, Calico scientists G. Ruby, M. Smith, and R. Buffenstein demonstrated that the long-lived naked mole rat has a flat mortality curve, defying Gompertzian laws[212] that an animal's risk of death grows exponentially with age. Their research has shown that the chance of dying for mammals such as humans, horses, and mice, among others, increases exponentially with age. However, naked mole rats do not age in the same manner as other mammals. They show little to no signs of aging and their risk of death does not increase even at 25 times past their time to reproductive maturity.[213]

Through sequencing, we found the naked mole rat's genome size is approximately 2.66 Gb, and 22,561 coding genes have been annotated. Compared with other mammalian genomes, the lost and acquired genes were 320 and 750 respectively. Among them, most of the missing genes are ribosome and nucleoside synthesis related genes. Some of the missing genes may be related to the deterioration of their vision.

In conclusion, naked mole rats are exceptional animals. Studying them will further our understanding of both cancer prevention and the biological mechanisms of longevity. The final question that remains is if humans really would like to extend life and reduce cancer risk by being as miserable as mole rats, by living naked in the dark and eating the queen's poop?

The Jellyfish (Scyphozoa, etc.)

Jellyfish are mainly free-swimming marine animals with umbrella-shaped bells and trailing tentacles, as shown in Fig. 6B.4.

Figure 6B.4 Jellyfish in medusa phase with umbrella bells and tentacles (source: Wikipedia).

The English name of the jelly fish is confusing because the jellyfish is not a fish, it is an invertebrate. Jellyfish have no bones, no brains, no hearts, and most of them are transparent. Although frequently depicted in their medusa phase as in Fig. 6B.4, not all jellyfish are free-swimming: some jellyfish are anchored to the seabed by stalks. The bell can pulsate to provide propulsion and highly efficient locomotion. The tentacles are armed with stinging cells and may be used to capture prey and defend against predators.

Jellyfish have a complex life cycle, shown in Fig. 6B.5.

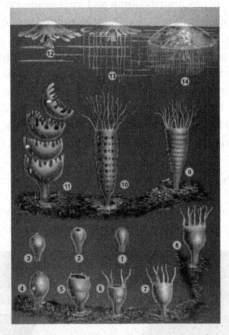

Figure 6B.5 Jellyfish life cycle. 1–3: Larva searches for site. 4–8: Polyp grows. 9–11: Polyp strobilates. 12–14: medusa grows into Scyphozoa (source: Wikipedia).

Jellyfish sperms fertilize eggs, which develop into larval planulae, become polyps, strobilate and then transform into adult medusae. During *strobilation*, the polyp's tentacles are reabsorbed and transverse constrictions appear in the upper part. The medusa is normally the sexual phase.

Jellyfish are mostly either male or female, or sometimes hermaphrodites. They release sperm and eggs into the surrounding water, where the unprotected eggs are fertilized and develop into larvae. The sperm may also swim into the female's mouth, fertilizing the eggs within her body.

Jellyfish get oxygen from the surrounding sea water through thin skins without breathing system. Although they look transparent, they have three layers of skin: the epidermis, the gastric cortex, and the mesoglobin. The jellyfish's body contains 95–98% water.

There is a general human feeling that jellyfish have always lived in an unknown sea, that they are transparent and soft, weak, incompetent, drifting with the current, etc. In fact, they have many secrets: many jellyfish are highly toxic; some can make people die in a short time. In fact, every year a large number of humans die from jellyfish attacks worldwide. Jellyfish can survive long periods of hypoxia, and they may even survive and benefit from urban sewage. Acid rain can promote jellyfish growth. Jellyfish can spawn quickly when attacked: therefore, killing a wild jellyfish sometimes results in the emergence of 100 small jellyfish. A large jellyfish population can exceed 100,000 members.

Jellyfish range in size from about 1 mm in bell height and diameter to nearly 2 m; the tentacles and mouth parts usually extend beyond this bell dimension. The smallest jellyfish are the peculiar creeping jellyfish, which have bell disks from 0.5 mm to a few millimeters in diameter, with short tentacles that extend out beyond this, used by these jellyfish to move across the surface of seaweed or the bottoms of rocky pools. Many of these tiny creeping jellyfish cannot be seen in the field without a hand lens or microscope; they can reproduce asexually by splitting in half (called fission).

The lion's mane jellyfish (*Cyanea capillata*) is one of the largest species of jellyfish, and arguably the longest animal in the world, with fine, thread-like tentacles that may extend up to more than 35 m. They have a moderately painful, but rarely fatal, sting. The increasingly common giant Nomura's jellyfish, found in some waters of Japan, Korea, and China in summer and autumn, is another candidate for "largest jellyfish," in terms of diameter and weight: the largest Nomura's jellyfish in late autumn can reach 2 m in bell (body) diameter and about 200 kg in weight. The large bell mass of the giant Nomura's jellyfish can easily dwarf a diver.

The impressive-looking deep-sea jellyfish *Stygiomedusa gigantea* is another candidate for the "largest jellyfish," with its thick, massive bell up to 1 m wide, and four thick, "strap-like" oral arms extending up to 6 m in length.

Most jellyfish have quite short life cycles, ranging from days to years, and some small jellyfish live only a few hours. Although jellyfish live for only a few weeks, they appeared on Earth very long ago—even earlier than the dinosaurs.

Human research based on jellyfish has made remarkable discoveries: the fluorescent protein in multitubular jellyfish can track cancer cells in the human body, and is also used to observe the growth of cancer cells. Roger Yonchien Tsien shared the 2008 Nobel Prize in chemistry with Osamu Shimomura and Martin Chalfie[214] for research on a glowing jellyfish protein, known as green fluorescent protein (GFP), that has revolutionized our ability to study disease and normal development in living organisms.

Another connection to human health may still be in doubt, although the dietary supplement Prevagen has become a big seller in the US. The manufacturer[215] claims that Prevagen supports a "healthier brain, sharper mind and clearer thinking." Prevagen contains apoaequorin, a protein that was first obtained in the 1960s from a specific type of jellyfish that glows. When apoaequorin is exposed to calcium, the protein and calcium bind and a blue light is produced. For more than 40 years, apoaequorin has been used to study how calcium works inside cells. Problems with calcium regulation in the human brain are thought to play a role in age-related mental decline. Because apoaequorin has a similar structure to human calcium-binding proteins, the manufacturers of Prevagen believe it might help regulate calcium in the brain and thereby reduce memory loss and mental decline. However, the efficacy of this dietary supplement has not yet been fully confirmed and recognized by the FDA.

In 2013, *Science* magazine[216] released a genome of the *Mnemiopsis leidyi* (a kind of jellyfish). The genome is very small, only 150M. On the basis of sequencing data, the researchers pointed out that Ctenophores[e] were older than other organisms with complex nervous systems, meaning that the nervous system evolved independently twice.

In 2014, *Nature* magazine[217] published a draft genome of *Pleurobrachia bachei* and the transcriptome of 10 other Ctenophores

---

[e]An aquatic invertebrate of the phylum *Ctenophora*, which comprises the comb jellies that swim by beating their combs rhythmically to push themselves forward.

species. The scientists found that the genome sequence omitted genes that are common in all other animals, including normal genes related to immunity, development, and nerve function. In view of this situation, researchers believe that Ctenophores evolved an independent nervous system. In this study, the researchers found that the neural, immune and developmental gene content of these genomes was significantly different from that of other animal genomes, such as the absence of HOX genes and standard microRNA mechanisms, and the reduction of immune gene replenishment. Moreover, many bilateral neuron-specific genes and "classical" neurotransmitter channel genes do not exist or are not expressed in neurons. Thus, the scientists suggest that the nervous system (and possibly muscle differentiation) of Ctenophores evolved independently.

Although scientists have long believed that human evolution started out initially from sponges, new genetic studies do suggest otherwise. Sponges are simple creatures, lacking muscles or a nervous system, while comb jellies have both. Those gelatinous marine animals that look similar to jellyfish may actually have been the first animals to have evolved over 600 million years ago.[218]

The Cone Snail (Conus magus, etc.)

Cone snails are a group of very nice-looking but extremely venomous predatory sea snails, shown in Fig. 6B.6.

There are hundreds of different species of cone snails. They are typically found in warm and tropical seas and oceans worldwide, in particular in the Western Indo-Pacific region. Some species are adapted to more temperate environments, such as the Cape coast of South Africa, the Mediterranean, or the cool waters of southern California. They live on sand or among rocks or coral reefs. When living on sand, cone snails tend to bury themselves with only the siphon protruding from the surface.

Cone snails are carnivorous and hunt and eat prey such as marine worms, small fish, mollusks, and even other cone snails. They are known for their venom attacks on faster-moving prey, such as fish. The venom of a few larger cone snail species is powerful enough to kill a human being.

Figure 6B.6 Cone snails have beautiful patterns. Their shells are collectors' items (source: Wikipedia).

The sensory modality that cone snails use to become aware of the presence of a prey animal is called *osphradium*, an olfactory organ that is also linked with the respiration organ. This organ is thought to be testing incoming water for silt and possible food particles. It also relies on chemoreception (not vision) for prey detection. The cone snails immobilize their prey using a modified, dartlike, barbed radular tooth, made of chitin,[f] along with a venom gland containing neurotoxins. The needle-like modified tooth is often compared to a harpoon. It is barbed and can be extended some distance out from the head of the snail. Small species of these cone snails hunt small prey, such as marine worms, whereas larger cone snails hunt fish.

Their venomous sting will occur without warning and can be fatal. The species most dangerous to humans are the larger ones that prey on small bottom-dwelling fish. Cone snail venom is also showing great promise as a source of new drugs.[219] It contains an impressive diversity of chemical compounds, collectively numbering over 50,000 small, disulfide-rich peptides.

---

[f]Chitin, similar to cellulose in plants, is the main ingredient in the exoskeletons of crustaceans. Crustaceans are an arthropods taxon that includes animals such as crabs, lobsters, crayfish, shrimp, etc.

Cone snail venom is fast-acting: bad news for fish but potentially good news for people with diabetes. These marine organisms produce "weaponized" insulin that may be more effective than traditional drugs that treat hyperglycemia. In a new study published in *Nature Structures & Molecular Biology*,[220] it was found to be faster and more effective than the best rapid response insulin currently used by humans, and will begin to function within five minutes of entering the body. In comparison, the fastest insulin currently used by humans takes fifteen minutes. The study also found that the maximum difference between cone snail insulin and current genetically engineered human insulin is that it lacks a molecular segment that allows it to stick to other insulins. Therefore, cone snail insulin is not working as well as human insulin. However, the fact that it acts quickly may still provide an opportunity for human use as an alternative.

As stated in a recent review,[221] the cone snail venom peptides have become a valuable source of neuro pharmacological targets as many of them *selectively* modulate ion channels and transporters.

Already, there is an FDA approved (2004) pain medicine, Ziconotide, that is much more powerful than morphine. It was developed by Elan and Eisai and is marketed as the drug Prialt.[222] Its only drawback is that it has to be injected into the spinal fluid to be effective.

Other cone snail venom derived drug candidates are in clinical and preclinical trials, for indications such as the treatment of heart disease and stroke, Alzheimer's disease, Parkinson's disease, depression, and epilepsy.

In 2010, AstraZeneca's biotech division MedImmune signed an agreement[223] with Xenome Biotech, based in Queensland, Australia, to develop pain medicine based on cone snail venom. MedImmune has been screening Xenome's library of 2000 compounds for drug candidates in the area of pain management for over a year and has now obtained exclusive licenses for four peptides. The Australian Great Barrier Reef provides an ideal environment for cone snails (Fig. 6B.7) and Xenome has focused on the extraction of peptides from cone snail venom tissue and use them to inhibit similar nerve pathways in humans.

Figure 6B.7   Cone snail in marine environment (source: Wikipedia).

The Komodo Dragon (Varanus komodoensis)

The Komodo dragon is the world's largest lizard and can be found in the five Indonesian islands of Komodo, Rinca, Flores, Gili Motang, and Padar.

According to fossil records, Komodo dragons moved out of Australia and made their way to the Indonesian islands, arriving around 900,000 years ago.

As a result of their size, these lizards dominate the ecosystems in which they live. Komodo dragons hunt and ambush prey including invertebrates, birds, and mammals. It has been claimed that they have a venomous bite: indeed, there are two glands in the lower jaw which eject several toxic proteins, rapidly decreasing the prey's blood pressure, causing extreme blood loss, tissue damage and pain. A bite will send victims as big as buffalos into shock that will render it too weak to fight. In the venom, some compounds that reduce blood pressure are as potent as those found in the world's most venomous snake, western Australia's inland Taipan. Komodo dragons are also known to hunt in groups, an exceptional behavior among reptiles. The diet of big Komodo dragons mainly consists of Timor deer (Fig. 6B.8), though they also eat considerable amounts of carrion.

Figure 6B.8 Komodo dragons in Indonesia, and a "dragon" hunting deer (source: Wikipedia).

Komodo dragons have bifurcated tongues that recognize subtle odors. As long as prey is exposed to the air, they will find it quickly and kill it. They also tend to insert their tongues into the prey's mouth.

Komodo dragons eat by tearing large chunks of flesh and swallowing them whole while holding the carcass down with their forelegs. For smaller prey up to the size of a goat, they are able to swallow the whole prey whole. They use large amounts of red saliva to lubricate the food, but still need a long time (15–20 minutes) to swallow a goat. After eating up to 80% of its body weight in one meal, the dragon drags itself to a sunny location. Because of their slow metabolism, large dragons can survive on as few as 12 meals a year. Komodo dragons are a scary sight and are said to have inspired horror movie makers.

Komodo dragons are further able to perform *parthenogenesis*, something only a few other reptiles (such as python snakes) are able to do: females can produce offspring based on eggs that were generated without a male's participation, by fusing one egg cell with another cell to obtain a complete set of chromosomes.

What makes Komodo dragons particularly interesting in connection with human health is their ability to eat carrion without suffering from associated illnesses: they carry antimicrobial peptides as infection defense.

They endure numerous strains of pathogenic bacteria in their saliva and easily recover from wounds inflicted by other dragons, reflecting an exceptional innate immune defense. Scientists[224] from George Mason University (Manassas, VA), Virginia Polytechnic Institute (Blacksburgh, VA), and the University of Florida (Gainesville, FL), have combined partial de novo peptide sequencing with transcriptome assembly to identify antimicrobial peptides from Komodo dragon blood plasma. They identified 48 novel potential antimicrobial peptides. All but one of the identified peptides were derived from histone proteins, defined as highly alkaline proteins found in eukaryotic cell nuclei. They are the chief protein components of chromatin, acting as "spools" around which DNA winds. Without histones, the unwound DNA in chromosomes would be very long: for example, each human diploid cell (containing 23 pairs of chromosomes) has about 1.8 m of (unwound) DNA. In its "wound" state, however, the diploid cell contains only about 90 μm (0.09 mm) of chromatin. Histones are also playing a role in gene regulation.

The antimicrobial effectiveness of eight of these Komodo dragon specific peptides was evaluated against *Pseudomonas aeruginosa* (ATCC[g] 9027) and *Staphylococcus aureus* (ATCC 25923), with seven peptides exhibiting antimicrobial activity against *both* microbes, and one only showing significant potency against *P. aeruginosa*. This study demonstrates the power and promise of the scientific team's "bioprospecting" approach to antimicrobial peptide discovery, and it proves the presence of novel histone-derived antimicrobial peptides in the plasma of the Komodo dragon. These findings confirm the role that intact histones and histone-derived peptides play in defending the host from infection.

In addition, it has been demonstrated that a short peptide named DRGN-1 has the added observed benefit of significantly promoting wound healing in both uninfected and mixed biofilm[h] infected wounds.[225] It may have therapeutic potential as a topical compound for the healing of infected wounds.

---

[g]ATCC (American Type Culture Collection) is a nonprofit organization that collects, stores, and distributes standard reference microorganisms, cell lines, and other materials for R&D.

[h]A thin, slimy film (or layer) of bacteria that adheres to a surface and is difficult to remove. Biofilms protect bacteria during infection and support microbial survival.

The Gila Monster (Heloderma suspectum) *and Salamander* (Urodela)

The Gila monster (Fig. 6B.9) is a species of venomous lizard native to the southwestern United States and northwestern Mexican state of Sonora. A heavy, typically slow-moving lizard, up to 60 cm long, the Gila monster is one of only two known species of venomous lizards in North America, the other being its close relative, the Mexican beaded lizard. Although the Gila monster is venomous, its sluggish nature means it represents little threat to humans. However, its bite is very painful and it has acquired a fearsome reputation. It is sometimes killed despite being a protected species.

Figure 6B.9   Head of Gila monster and spotted salamander (source: Wikipedia).

The name "Gila" refers to the Gila River Basin in the states of New Mexico and Arizona. Heloderma means "studded skin" in Greek. Finally, "suspectum" was chosen by the paleontologist Edward D. Cope, because he "suspected" the lizard to be venomous.

More than a dozen peptides and other substances have been isolated from the Gila monster's venomous saliva. Four potentially lethal toxins include horridum venom, which causes hemorrhage in internal organs and exophthalmos (bulging of the eyes), and helothermine, which causes lethargy and partial paralysis of the limbs.

Most are similar in form to vasoactive intestinal peptide (VIP), which relaxes smooth muscle and regulates water and electrolyte secretion between the small and large intestines. The constituents of the lizard's venom that have received the most attention from researchers are the bioactive peptides, including helodermin, helospectin, exendin-3, and exendin-4. Helodermin has been shown to inhibit the growth of lung cancer.[226]

It turns out that the Gila monster's peptide *exendin-4* is similar to the GLP-1 hormone in the human digestive tract. GLP-1, glucagon-like peptide-1 analogue, increases the production of insulin when blood sugar levels are high. Insulin then helps move sugar from the blood into other body tissues where it is used for energy. The lizard hormone remains effective much longer than the (50% similar) human hormone, and thus its synthetic form helps diabetics keep their blood sugar levels from getting too high. The drug Exenatide, which is based on exendin-4's synthetic form, stimulates insulin production and also slows the emptying of the stomach, thereby causing a decrease in appetite that leads to weight loss.

The additional benefit of weight loss makes Exenatide a great choice for type 2 diabetes patients that are overweight.[227] Byetta (Exenatide's trade name) is manufactured by Amylin Pharmaceuticals Inc. in collaboration with Eli Lilly and Company. It comes in a prefilled pen that type 2 diabetics use to give themselves twice-daily injections within an hour before their morning and evening meals

Byetta, was approved by the Food and Drug Administration in April 2005 to treat type 2 diabetes in patients who were not able to get their high blood sugar under control in a combination with one or more of three other medications such as metformin, sulfonylureas, and thiazolidinediones.

Weight loss was not the only significant finding. After three years of including exenatide in the drug regimen, 46% of participants achieved sustained blood-sugar levels considered healthy by the American Diabetes Association (ADA).

Byetta even trumped the Lantus insulin treatment according to DURATION-3 trial results presented at the annual American Diabetes Association (ADA) conference in Orlando, Fla., and simultaneously published in the June 24, 2010, edition of the *Lancet*.[228]

The Salamander

Salamanders (see right half of Fig. 6B.9) are a group of amphibians characterized by a lizard-like appearance. All present-day salamander families are grouped together under the order Urodela. They have permeable skin and live in habitats in or near water. Some salamander species are fully aquatic throughout their lives, some take to the water intermittently, and others are entirely terrestrial as adults.

They are known to regenerate lost limbs, as well as other damaged parts of their bodies. Scientists focused on regenerative medicine hope to reverse-engineer the salamander's remarkable regenerative processes for potential human medical applications, such as brain and spinal cord injury treatment or preventing harmful scarring during heart surgery recovery.[229]

The molecular basis of regeneration has been investigated[230] by a British team led by Jeremy Brockes, and it has been found that the surface protein Prod 1 and anterior gradient protein family member nAG are critical. Protein nAG is a secreted ligand for Prod 1, and a growth factor for cultured salamander blastemal cells. Protein nAG is sequentially expressed after amputation in the regenerating nerve and the wound epidermis, both key tissues of the stem cell niche. Denervation interrupts the regenerative process. However, the local expression of nAG after electroporation is sufficient to rescue a denervated blastema and regenerate the structures.

What had been known already before this study was that there are key tissues of limb regeneration, namely the regenerating peripheral nerves and the wound epidermis. The severed axons retract after amputation, and then regenerate back along the nerve sheath and into the blastema. Axonal regeneration can be prevented or arrested by transecting the spinal nerves at the base of the limb, distant from the amputation level. The generation of the initial cohort of blastemal cells occurs in a denervated limb, but the growth and division of these cells depends on the concomitant regeneration of peripheral axons. Both motor and sensory axons have this activity, and it is independent of impulse traffic or transmitter release. Limb regeneration is abrogated if the blastema is denervated during the initial phase of cellular accumulation, but denervation after the mid-bud stage allows the formation of a regenerate. The wound epidermis is not required to support proliferation during the first week of regeneration in an adult salamander, but it is critical for subsequent division. Immune cells called *macrophages* are extremely critical in the early stages of regenerating lost limbs. Wiping out these cells permanently prevented regeneration and led to tissue scarring!

Brockes et al. conclude with the following statement: "It is striking that the expression of a single protein can rescue limb regeneration in an adult animal. This finding underlines that the blastema is an autonomous unit of organization for which there is no obvious mammalian counterpart. We suggest that one approach for regenerative medicine would be to understand the specification of the blastema at a level of detail that would allow it to be engineered in mammals."

The skin of some species contains the powerful poison tetrodotoxin, a neurotoxin similar to the pufferfish venom. These salamanders tend to be slow-moving and have bright warning coloration to advertise their toxicity. Tetrodotoxin is the most toxic nonprotein substance known. Ingestion of even a minute fragment of skin is deadly. Tetrodotoxin is a sodium channel blocker. It inhibits the firing of action potentials in neurons by binding to the voltage-gated sodium channels in nerve cell membranes and blocking the passage of sodium ions (responsible for the rising phase of an action potential) into the neuron. This prevents the nervous system from carrying messages and thus muscles from flexing in response to nervous stimulation.

Legends have developed around the salamander over the centuries, many related to fire. When placed into a fire, salamanders dwelling inside a rotten log would attempt to escape from the log, lending to the belief that salamanders were created from flames. Pope Alexander III had a salamander tunic which he valued highly. The mythical fire salamander was said to be so toxic that by twining around a tree, it could poison the fruit and so kill any who ate them and by falling into a well, could kill all who drank from it.

Salamanders are also known for their gigantic genomes, spanning the range from 14 Gb to 120 Gb,[231] much larger than the human genome at only 3.2 Gb. The genomes of *Pleurodeles waltl* (20 Gb) and axolotl (*Ambystoma mexicanum*) (32 Gb) have been sequenced. The 32 billion base pair long sequence of the axolotl's genome was published in 2018 and is the largest animal genome completed so far. It revealed species-specific genetic pathways that may be responsible for limb regeneration.[232] Although the axolotl genome is about 10 times as large as the human genome, it encodes a similar number of known proteins, namely 23,251.

The axolotl is still used as a model organism, and large numbers are bred in captivity. They are especially easy to breed compared to other salamanders in their family. An attractive feature for research is the large and easily manipulated embryo. Axolotls are used in heart defect studies due to the presence of a mutant gene that causes heart failure in embryos. Since the embryos survive almost to hatching with no heart function, the defect is very observable. The axolotl is also considered an ideal animal model for the study of neural tube[i] closure due to the similarities between human and axolotl neural plate and tube formation.

The axolotl is capable of the regeneration of entire lost appendages and even more vital structures in a period of months. Some have even been found restoring the less vital parts of their brains. They can also readily accept transplants from other individuals, including eyes and parts of the brain—restoring these alien organs to full functionality.

The Burmese Python (Python bivittatus)

The Burmese Python (Fig. 6B.10) has recently attracted the attention of both scientists interested in metabolic diseases such as diabetes, and environmentalists concerned about Florida's Everglades National Park.

The scientists are seriously considering the possibility of treating this large snake as a "model organism,"[233] while the U.S. Department of the Interior has banned importation and Everglades park rangers are observing that raccoon populations are down by 99.3%, opossums by 98.9%, and white-tailed deer by 94.1%.[234] The Burmese pythons in the Florida everglades have reached enough critical mass to be considered a "viable population" and an invasive species. As the story goes, Hurricane Andrew in 1992 destroyed a python breeding facility and the escaped snakes spread and populated the Everglades. Also, during the 10 years 1996–2006,

---

[i]The neural tube is the embryonic structure that ultimately forms the brain and spinal cord.

more than 90,000 Burmese pythons were imported into the US as "pets."

Figure 6B.10   The Burmese python as photographed in the Florida Everglades (source: Wikipedia).

Like all snakes, the Burmese python is an elongate carnivorous reptile that has no legs and belongs in the suborder serpents. Snakes and other reptiles, along with birds, are regarded as having a single evolutionary lineage and at one point in the past shared a common ancestor. Snakes probably descended from a lineage of fully legged lizards.

Snakes are identified by the complete lack of eyelids and external ears. There are about 2900 known species of snakes, of these, some 700 species are venomous and of those around 250 can deliver a bite capable of killing a human. But even nonvenomous snakes can be dangerous to humans, especially the biggest snakes in the world. The Burmese python is definitely among the biggest. In the wild, they average close to 4 m and can reach close to 6 m in length. Females are both longer and bulkier, hence heavier, than males.

Snakes inspire fascination and feelings in a way that no other type of animal can. Along with spiders, leeches and other "creepy crawlies," people often perceive snakes as animals to fear and hate. Snakes first appeared on Earth about 100–130 million years ago. Snakes eat their prey whole and are able to consume prey up to three times larger than the diameter of their head. This is possible because their lower jaw and upper jaw can separate from each other. To keep prey from being able to escape, most snakes have rear-facing teeth that hold their victim in their mouths.

While venomous snakes inject their prey with venom (see pit viper chapter below), pythons are constrictors that will squeeze the life out of their prey. Snakes don't need to hunt and feed on a daily basis. They can survive for several months without food after feeding. They are able to reduce the metabolic rate by 70%. However, they will need water. Most snakes will hunt mainly at night. To sense the heat of warm-blooded animals nearby, the Burmese python has a series of labial pits on the sides of its lower jaws that can sense heat.

The lower jaw of a snake is only loosely attached to the skull, allowing snakes to open their mouths very wide. As an extra adaptation, the lower jaw bones of snakes are not joined together at the front. This allows each side of the lower jaw to move independently and helps the snake to stretch its mouth over large prey. Then the snake gradually "walks" each side of its jaws over the prey until it can be swallowed.

Snakes have no ears, and their visual abilities are poor. They have no nose, but they do have a strong sense of smell.

Burmese Pythons are excellent swimmers and need a permanent source of water. They can stay submerged for up to half an hour. They can be found in grasslands, marshes, swamps, river valleys, and jungles with open clearings. They are also good climbers.

Burmese pythons breed in the early spring, with females laying up to about 30 eggs in March or April. They wrap themselves around the eggs to raise the temperature around the eggs. They then wait until the hatchlings use their teeth to cut their way out of their eggs. After that, the baby snakes are left to themselves.

What makes the pythons so special and potentially important for human health is their digestive system. The digestive response of Burmese pythons to large prey has made them a model species for digestive physiology. A fasting python has a reduced stomach volume and acidity, reduced intestinal mass, and a "normal" heart volume. After ingesting prey, the entire digestive system undergoes a massive transformation, with rapid hypertrophy of the intestines, production of stomach acid, and a 40% increase in mass of the ventricle of the heart to fuel the digestive process.

In his NY Times Magazine article (May 2017) referenced above, Daniel Engber describes the animal as nothing but "a slithering digestive tract." A single meal for a grown python may contain more than 50,000 calories, but the python is fully capable to deal with this overload. Its liver, pancreas and kidneys double in mass and its insulin level increases significantly as its body temperature rises and its pulse triples.

What may be most noteworthy are changes in the snake's metabolic rate and how snake blood gets so clogged with fatty acids that it looks like milky white yogurt. It has been found that blood plasma after a python meal acts like a "cell-expanding cocktail," an explanation for the above-mentioned expansion of the snake's inner organs. It has been found that pouring snake blood on a dish of rat heart cells has a similar effect.

The metabolic rate or "speed of metabolism" is the number of calories one burns in a given amount of time. The basal metabolic rate (BMR) is defined as the metabolic rate at rest (or when asleep). Exercise increases the metabolic rate, and so does the intake of proteins. For humans, the biggest exercise induced metabolic rate increase—up to five times—has been recorded for bike racers in competitions like the Tour de France. The Burmese python, however, is capable of increasing the metabolic rate by a factor of up to 44 after a heavy meal! Even among snakes, that is very impressive. By comparison, rattlesnakes only increase it eightfold.

While we have not seen any approved therapeutic treatments as a result of Burmese Python research, several scientific projects have started at various places, such as the Broad Institute of Harvard and MIT, the University of Alabama,[235] the University of Colorado,[236] and the University of Texas at Austin.[237,238] They are all focused on metabolism. Clinical applications may be type 2 diabetes, Obesity, and cardiovascular disease.

Burmese pythons (as well as other snakes) massively downregulate their metabolic and physiological functions during extended periods of fasting. During this time their organs atrophy, saving energy. However, upon feeding, the size and function of these organs, along with their ability to generate energy, dramatically increase to accommodate digestion.

Judging from gene expression profiles before and after feeding, a few sets of genes are influencing the significant changes observed in the pythons' internal organ structure. Key proteins, produced and regulated by these important genes, activate a cascade of diverse, tissue-specific signals that lead to regenerative organ growth. Some genes that are expressed steadily in daily life become more active, churning out massive amounts of proteins that promote growth or metabolism. What is perhaps surprising is the strong response shown by mammalian cells to serum obtained from post-feeding pythons. The response suggests that the signaling function is conserved across species and could perhaps one day be used to improve human health. The hope is that our understanding of how snakes accomplish organ regeneration could be translated to the treatment of human metabolic and cardiovascular diseases, perhaps even cancer: for example, a gene called GAB1, which appears altered in the snake genome, is involved in some human cancers.

6C. Animals With Potential to Improve Human Performance

Sapiens has gained a lot of inspiration from animals. They have helped us to create new types of mechanical systems, instruments, building structures and even weapons.

Our superior brain and our flexible and fine-tuned hands have overcome many human challenges. When we admired birds that can fly, we first tried to attach wings to our arms. When those attempts failed, we developed more elaborate structures and eventually ended up building airplanes. When our eyes were unable to either explore the microscopic world or the nightly skies, we invented optical instruments such as microscopes or telescopes.

By observing animals, we developed ideas that could help us to defend ourselves or to attack the enemy in a time of war.

For example, by observing the **squid**, human scientists have found that the sacs in the squid can secrete a black liquid. When the squid encounters danger, it will spurt the black liquid, which will entice the enemy to chase it and let the squid escape. Inspired by the squid, human engineers invented the torpedo bait. The torpedo bait resembles a submarine, which will make the enemy unable to distinguish the true and false torpedo target and give time for the submarine to escape.

What did we learn from the **giraffe**, the world's tallest animal? It measures about three meters from head to heart, and pumping blood to the brain requires blood pressure of up to 160–260 mm Hg. Therefore, when a giraffe bows its head and drinks water, it should be suffering from severe high blood pressure. However, the giraffe is still alive and conscious. What is the reason? After careful investigation, scientists realized that the giraffe has evolved a thick layer of skin around the blood vessels, thereby limiting blood pressure. Aircraft designers and aero-biologists consequently designed a novel "anti-wear suit" based on the principle of giraffe, which solved the pain caused by brain ischemia during the sudden acceleration experienced by high-speed fighter pilots. There is a device in this anti-wear suit that compresses the air when the airplane accelerates, and also exerts corresponding pressure on the blood vessels.

Another example is the study of **wild boars** that has led to the development of gas masks. The world's first gas masks were based on the exceptional ability of the wild boar's nose to limit poisonous gases to penetrate!

In addition, researchers have developed a number of camouflage methods for the military by studying the discoloration ability of **chameleons**.

Scientists studied the eyes of **frogs** and invented electronic frog eyes, designed to differentiate between meaningful and less important stimuli.

The US Air Force has developed a miniature thermal sensor that emulates the "hot eye" function of the **viper**. It functions in the dark when eyes can no longer identify warm-blooded living objects passing by.

Even modern radar systems, which are radiolocation and ranging devices, are closely linked to animal capabilities: instead of using their eyes to navigate in the dark, **bats** have developed an echolocation system consisting of the mouth, throat, and ears. Because the bat emits ultrasonic waves while flying, it can detect the ultrasonic waves reflected by the obstacles. Inspired by this, human engineers have designed modern radar systems.

In this chapter we will introduce animals that could help us to improve our performance in more or less obvious ways. As a first step, we need to study how they utilize their exceptional capabilities. Then we need to figure out whether we need to build an instrument or machine to emulate them, or if there may be a "bionic" way to directly augment a human capability.

The Atlantic Bay Scallop (Argopecten irradians/*Pectinidae*)

Scallops are very interesting animals with strong human connections. In particular, the Atlantic bay scallop (Fig. 6C.1) has intrigued humans since ancient times.

The majority of scallop species live in saltwater on sandy substrates. When they sense the presence of a predator (such as a starfish or a sting ray), they may attempt to escape by swimming swiftly through the water, using jet propulsion. They create their somewhat erratic movement by repeatedly clapping their shells together.

Figure 6C.1 The Atlantic bay scallop (source: Wikipedia).

Scallops have a well-developed nervous system. Figure 6C.1 shows a ring of up to 200 simple eyes, situated (like a necklace) around the edge of their shells. We will return to the amazing capabilities of those "reflector telescope eyes" below.

Most connoisseurs of fine food know and appreciate scallops for their delicious taste, in particular when their adductor muscles are prepared with butter and white wine. They are appreciated as a delicacy around the world, in cuisines as diverse as French, Spanish, Japanese, and Chinese.

Figure 6C.2a shows the anatomy of a live opened scallop. The pale circular part is the adductor muscle (used to open and close the shells). The darker orange curved part is the "coral," a culinary term for the ovary or roe.

The shell is frequently used for decorative purposes and has been the subject of some famous works of art as discussed below. Figure 6C.2b shows a 3–5 million-year-old fossil.

As for reproduction, the scallop family includes dioecious members (i.e., males and females are separate), hermaphrodites (both sexes in the same individual), and the occasional protoandrous hermaphrodites, characterized by a transformation of males into females when they reach adulthood. Females have red roe while the roe of males is white. Spermatozoa and ova are released freely into the water during mating season, and fertilized ova sink to the bottom. After several weeks, the immature scallops hatch and the larvae, miniature transparent versions of the adults called spat, drift in the plankton until settling to the bottom. There they start growing, usually attached to a solid surface. There is rapid growth within the first several years, with a significant increase in shell size and quadrupling of meat weight. Scallops reach commercial size at about four to five years of age. Some scallops have been known to live over 20 years.

(a)   (b)

Figure 6C.2   (a) Scallop anatomy. (b) Scallop fossil: Nicosia formation (pliocene) of cyprus (source: Wikipedia).

The scallop shell is the traditional emblem of Saint James and is popular with pilgrims returning from the Way of Saint James (El Camino de Santiago), starting in France and ending at the apostle's shrine at Santiago de Compostela in Galicia, Spain. Medieval Christians would collect a scallop shell while at Compostela as evidence of having made the journey. The association of St. James with the scallop can most likely be traced to the legend that the apostle (named Jacques in French and Tiago in Spanish) once rescued a knight covered in scallops. The French name of the scallop is indeed *coquille St. Jacques*.

When referring to St. James, or to a completed pilgrimage, a scallop shell valve is shown with its convex or outer surface. It is a popular symbol for rescuing, healing, or resurrecting a dying knight, and has been incorporated in numerous coats of arms (e.g., Winston Churchill's).

On the other hand, when the scallop shell is used as a **fertility** symbol, it is displayed with its concave interior surface, as illustrated in Fig. 6C.3.

Figure 6C.3 Birth of Venus, 15th century, painted by Sandro Botticelli (source: Wikipedia).

Let's switch back now to biology and what could be learned from a closer study of the (Atlantic harbor) scallop's amazing eyes, as recently investigated by scientists at Israel's Weizmann Institute and Sweden's University of Lund.[239] Before their contribution, it was already known that scallop eyes do not rely on a lens, but rather on a number of concave, parabolic mirrors, consisting of guanine crystals, to focus and retro-reflect light. Guanine (G) (Fig. 6C.4) is one of the four nucleobases in DNA, but is an interesting molecule of life by itself.

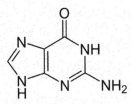

Figure 6C.4 Guanine $C_5H_5N_5O$ (source: Wikipedia).

G got its name from *guano*, excreta of seabirds from which guanine was first isolated. Guano was used as a source of fertilizers, a property already known to the Incas. It is said that the word *guano* actually has its roots in the Quechua language, still spoken in Peru. Crystalline guanine has been known for centuries as "pearl essence" and used in the cosmetics industry to provide a shimmering luster. Guanine crystals are rhombic platelets composed of multiple transparent layers. They have a high index of refraction that partially reflects and transmits light from layer to layer, thus producing a pearly luster.

Guanine is also found in specialized skin cells of fish, responsible for the silvery tint of scales. The well-known skin color changes of chameleons can also be explained by the way guanine crystals reflect and absorb light.

The recent progress in our understanding of scallop vision has been largely based on a new experimental technique, cryogenic scanning electron microscopy (cryo-SEM, recently awarded with the Nobel Prize for chemistry[240]). By imaging the ultrastructural properties of a single scallop eye mirror, B. Palmer et al found that a multilayered mirror is constructed from 20 to 30 layers of guanine crystals, separated by thin layers of cytoplasm. Each layer is composed of a tiling of closely packed crystal plates (Fig. 6C.5) organized so that each **square** crystal lies directly beneath another crystal in an adjacent layer, forming a vertical stack.

Figure 6C.5 Square guanine crystals in scallop eyes (see Science 01, December 2017, Vol. 358, Issue 6367, pp. 1172–1175. (Reprinted with permission from AAAS and from Palmer et al., Weizmann Institute, Israel).

The observed high precision suggests that crystal growth is strongly controlled. The crystals are arranged so that the high refractive index faces are oriented toward the direction of the incident light across the mirror, thereby creating a refractive mirror. This "crystal tiling" also minimizes surface defects and enables the formation of a highly reflective surface. Nature has created an optical device with guanine crystal building blocks of about 74 nanometer thickness and 86 nm cytoplasm spacing. The layer structure of the mirror is perfectly suited to reflect all the blue/green light that penetrates the scallop's habitat, and that the photoreceptors of both scallop retinas also effectively absorb wavelengths in the range of 475–540 nm. The mirror's nonspherical symmetry and tilt produce optical properties that are more complex than what would have resulted from a perfectly spherical shape: due to the mirror's shape, the best image formed on the "proximal" retina is at the periphery of the field of view, and the best image on the "distal" retina is formed just about the central field of view. The authors suggest that the scallop's well-focused peripheral vision is used to control and guide the scallop's movement while swimming. On the other hand, it is believed that the distal retina's function is related to relatively dark, moving features, triggering defense or escape reflexes.

The scientific team also created a computer model of the entire eye, permitting "ray-tracing" of the paths taken by light as it bounces off the mirrors. The requirements of "underwater vision" are different and the scallop's eyes have evolved a biological optical system that has to be admired. The final step, then, is the processing and coordination of the visual signals created by those refractive mirrors, by means of a cluster of neurons. The scallop may not have a brain but it definitely has an advanced "space age" visual system, optimized for underwater habitats.

Is there an opportunity to create bionic vision? Is there something humans could learn and apply? Could those stacked and square guanine crystals inspire the engineering of new materials with well-defined optical properties?

To start with, however, we may first need to understand the genetic characteristics and associated mechanisms that are responsible for the formation of those quite amazing (Newtonian) reflector telescopes that form the basis of the scallop's vision.

The Owl (Strigiformes)

Owls have been admired and feared for centuries by humans the world over. They have been roaming the planet Earth (except Antarctica) for about 60 million years, shortly after the extinction of the dinosaurs. They are featured in many books and works of art, mostly because of their special looks and mysterious behavior that includes haunting "hoot-hoot" calls. These birds possess a combination of unique qualities that make them successful hunters.

Their charismatic round faces (Fig. 6C.6) make them easily recognizable, and frequently they are chosen as symbols for book learning and wisdom. They are birds of prey, similar to hawks, eagles, and falcons, but they also hunt effectively at night when they need more than superior vision. According to research performed by contributors to the US Public Broadcasting Service (PBS),[241] they have developed extraordinary "superpowers" in the areas of vision, hearing, and "stealth flying."

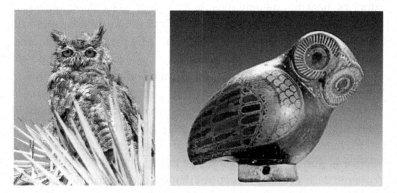

Figure 6C.6 Great horned owl and Greek Owl sculpture (640 BC) (source: Wikipedia).

Owls have especially large eyes and a high density of rod cells in their retinas. As already discussed above (human vision), rod cells are photoreceptor cells that can function in less intense light than cone cells. Cone cells render color; rods do not. Their high-density rods allow owls to gather enough available light to hunt at night. Compared to humans, they are able to detect an image two and a half times less bright. The frontal placement and binocular (far-sighted, unable to focus on objects nearby) function of owl eyes makes it difficult to move them. Instead of moving their eyes, owls therefore swivel their heads to view their surroundings. Their heads are capable of swiveling through an angle of roughly 270°, easily enabling them to see behind them without moving the body. Owls have 14 neck vertebrae compared to seven in humans, leading to increased flexibility. In addition, they are equipped with adaptations to their circulatory systems, permitting rotation without cutting off blood to the brain, and ensuring blood supply while they rotate their necks.[242] This ability to keep bodily movement at a minimum further reduces the noise the owl makes as it waits for its prey.

Amazingly, owls are also capable of hunting without even seeing their prey. A blanket of snow may deter other birds of prey, but an owl can hear its target (frequently a rodent) underneath. The owl's whole head is designed for superior hearing: its distinctive round face is shaped specifically to detect sound. Again, compared to humans, owls are capable to hear better by a factor up to 10×, and across a wider frequency range. Here is another reason for the owl's superior hearing, namely asymmetrical ear placement. With ears set at different places on its skull, an owl is able to determine the direction from which the sound is coming by the minute difference in time that it takes for the sound waves to reach the left and right ears. The owl will turn its head until the sound reaches both ears at the same time, at which point it is directly facing the source of the sound. This time difference between ears is a matter of about 30 millionth of a second, but it can be detected by the owl.

Perhaps the most amazing trait of owls is their silent flight. The silence is due to several factors including special "flight feathers" (covered with a velvety structure) that reduce air turbulence around the wing. This stealth-like power is another reason why owls are such successful hunters. Figure 6C.7 further illustrates this fact.

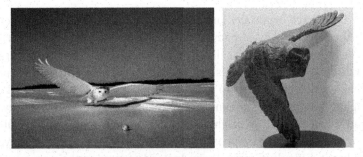

Figure 6C.7   The snowy owl in flight and sculpture illustrating flight of the great grey owl (source: Wikipedia).

An owl's sharp beak and powerful talons allow it to kill its prey before swallowing it whole. Scientists studying the diets of owls are helped by their habit of regurgitating the indigestible parts of their prey (such as bones, scales, and fur) in the form of pellets. These "owl pellets" ("spitballs") are easy to find and interpret, and are often used by schools for dissection by students of biology and ecology.

What could we learn from owls?

By studying their superior night vision, we could possibly help humans with impaired eyesight to regain some night vision and peripheral vision. It's not only about lost rods and cones of the retina (see Fig. 6C.8), there is also an important component residing in our respective brains that processes the information transmitted.

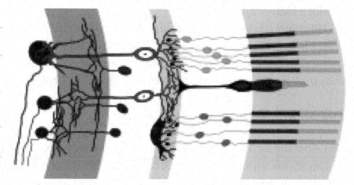

Figure 6C.8   Cross section of the retina. Rods are visible at far right (source: Wikipedia).

In addition, the development of "bionic eyes" could benefit from a detailed understanding of the owl's vision. Bionic eyes have been developed and tested since the late 1980s,[243] and are often based on retinal implants[244,245] connected to a camera, usually fitted to eyeglasses. Images are converted into electrical signals that are transmitted to the brain. Picture quality is still lacking but expected to improve steadily.

The devices consist of micro-electrodes surgically placed in or near one eye, along the optic nerve (which transmits impulses from the eye to the brain), or in the brain. The microelectrodes stimulate the parts of the visual system that is still functional in someone who has lost their sight. Electrical stimulation of the surviving neurons leads the person to perceive small spots of light. The optic nerve then transmits impulses to the brain from the retina at the back of the eye.

Several bionic eye projects are ongoing, including a project[246] by the University of Melbourne, Australia, and at least three approved commercial products, namely

Argus II[247] developed in the USA,

Alpha-AMS[248] in Germany, and

IRIS V2[249] in France.

To improve image quality, one may increase the number of implanted micro-electrodes and make them smaller. New "nanomaterials" may allow the electrodes to be small enough to produce high-acuity resolution.

Ultimately, scientists are seeking to fully understand the way our retina is communicating with the brain. To transmit a correct and "natural" visual message to the brain, we need to understand and then replicate the neural firing patterns of photoreceptors.

The Pit Vipers (Crotalinae)

The Crotalinae, commonly known as pit vipers, are a subfamily of venomous vipers found in Eurasia and the Americas. They are distinguished by the presence of a heat-sensing pit organ located between the eye and the nostril on both sides of the head. These are also the only viperids found in the Americas and include rattlesnakes, lanceheads, and Asian pit vipers. The timber rattlesnake is a subfamily shown in Fig. 6C.9.

Figure 6C.9 Timber rattlesnake with infrared detecting pit between eye and nostril (source: Wikipedia).

The pit vipers range in size from the hump-nosed viper, with an average total length of only 30–45 cm, to the *bushmaster*, a species known to reach a maximum total length of about 3.6 m.

Most pit viper species (including most rattlesnakes) congregate in sheltered areas or "dens" to brumate (see Chapter 3), to benefit from the combined heat. In cool temperatures and while pregnant, they like to bask on sunny ledges.

All pit vipers share a deep depression or "pit" in the area between the eye and the nostril on either side of the head. These pits are the external openings to a pair of extremely sensitive infrared-detecting organs. Temperature sensors indirectly detect infrared radiation by its heating effect on the skin inside the pit. They can detect both the intensity and the direction of the heat source, usually emanating from a warm-blooded prey animal. By combining information from both pits, the snake can also estimate the distance of the object.

The heat pit consists of a deep pocket with a membrane stretched across it. Behind the membrane, an air-filled chamber provides air contact on either side of the membrane. The pit membrane is highly vascular and innervated with numerous heat-sensitive receptors formed from terminal masses of the trigeminal nerve (terminal nerve masses, or TNMs). The receptors are therefore not discrete cells, but a part of the trigeminal nerve itself. The purpose of the vasculature is to provide oxygen to the receptor terminals, and to rapidly cool the receptors to their thermo-neutral state after being heated by thermal radiation from a stimulus. Were it not for this vasculature, the receptor would remain in a warm state after being exposed to a warm stimulus, and would present the animal with "afterimages" even after the stimulus was removed.

The end result is that the pits act like a *sixth sense* of *thermoception* to help the snakes find and size the warm-blooded prey on which they feed. Most pit vipers are nocturnal, preferring to avoid high daytime temperatures. The snakes' heat-sensitive pits are also thought to help vipers in locating cooler areas where they may prefer to rest.

Even when deprived of their senses of sight and smell, these snakes can strike accurately at moving objects less than 0.2°C warmer than the background. The paired pit organs provide the snake with "thermal rangefinder" capabilities. These organs are of great value to a predator that hunts at night, as well as for avoiding the snake's own predators.

The ability to sense infrared thermal radiation evolved independently in several different families of snakes. Essentially, it allows these animals to "see" radiant heat at wavelengths between

5 and 30 μm (10^{-6} m) to a degree of accuracy such that a blind rattlesnake can target vulnerable body parts of the prey at which it strikes. In addition to prey detection, the pit organs are also used in thermoregulation and predator detection, making it a quite general-purpose sensory organ.

The facial pit actually visualizes thermal radiation using the same optical principles as a pinhole camera. The location of a source of thermal radiation is determined by the membrane of the heat pit. Although computer analysis is suggesting that the spatial resolution is quite poor, it is possible that the integration of visual and infrared perception may be used to help sharpen the image. In addition, snakes may deliberately choose ambush sites with low thermal background radiation (colder areas) to maximize the contrast of their warm prey in order to achieve a higher degree of accuracy from their thermal "vision."[250]

Pit vipers are also unique in that they have a specialized muscle, called the muscularis pterigoidius glandulae, that contracts to force venom out of the gland. Viper venom acts mostly on the vascular system, bringing about coagulation of the blood and clotting of the pulmonary arteries; its action on the nervous system is not great, no individual group of nerve-cells appears to be picked out, but there is some effect upon respiration. There is a considerable influence upon the circulation. The pain of the wound is severe, and is speedily followed by swelling and discoloration.

In conclusion, Pit vipers have developed the most accurate thermoception sense among all snakes. Although boas and pythons have pits, they lack the suspended membrane and have simpler anatomical structures in their pit holes.

The Vampire Bat (Desmodus rotundus)

Vampire bats are bats that rely only on blood as their food source, a dietary trait called *hematophagy*. Along with the hairy-legged vampire bat (*Diphylla ecaudata*), and the white-winged vampire bat (*Diaemus youngi*), the common vampire bat (Fig. 6C.10) is the only mammalian species that feeds solely on blood. All three vampire bat species are native to the Americas, ranging from Mexico to Brazil, Chile, Uruguay, and Argentina. The common vampire bat prefers warm and humid climates and forages in tropical and subtropical woodlands and open grasslands. They live in trees, caves, abandoned buildings, old wells, and mines.

Figure 6C.10 Common vampire bat and taxidermied vampire bat showing wings (source: Wikipedia).

Dominant adult bat males defend groups of females. The common vampire bat is one of the most social of bat species: it practices food sharing and social grooming, a behavior where animals clean or maintain one another's body or appearance. Because it feeds primarily on mammalian blood, particularly that of livestock such as cattle and horses, the common vampire bat is considered a "pest."

Vampire bats may be carriers of the rabies virus. When they do, they appear to be clumsy, disoriented, and unable to fly, behaviors that make it more likely that they will come into contact with humans. There is evidence that it is possible for the rabies virus to infect a host purely through airborne transmission, without direct physical contact of the victim with the bat.

Common vampire bats have good eyesight. They are able to distinguish different optical patterns and may use vision for long-range orientation. These bats also have well-developed senses of smell and hearing: the cochlea is highly sensitive to low-frequency acoustics, and the nasal passages are relatively large. They emit echolocation signals orally. Hence, they fly with their mouths open for navigation. They emit calls and listen to the echoes of those calls that return from various objects near them. As shown in Fig. 6C.11, they then use these echoes to locate and identify the objects. Their auditory systems are adapted for this purpose. They are highly specialized for sensing and interpreting the stereotyped echolocation calls characteristic of their fellow bats.

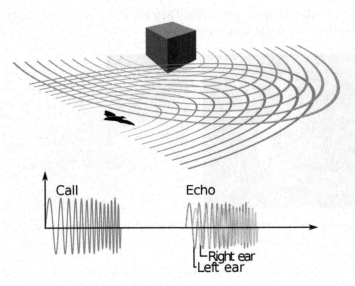

Figure 6C.11 Ultrasound signals emitted by a bat, and the echo from a nearby object (source: Wikipedia).

The common vampire bat is using both *echolocation* and *olfaction* to track down prey. They feed at distances of 5–8 km from their roosts. When a bat selects a target, it lands on it, or jumps up onto it from the ground, usually targeting the rump, flank, or neck of its prey; *heat sensors* in the nose help it to detect blood vessels near the surface of the skin. Trigeminal nerve fibers that innervate these IR-sensitive receptors are helping to detect the infrared thermal radiation emitted by their prey. Their infrared perception enables the vampire bats to localize warm-blooded animals within a range of about 10–15 cm. Having detected regions of maximal blood flow on targeted prey, the vampire bats then pierce the target's skin with their teeth, biting away a small flap, and then lap up the blood with their tongues. The blood is kept from clotting by an anticoagulant in the saliva.

Aside from "echolocation technology," there is another human connection to bats. In old folklore, a vampire is defined as "a being that subsists by feeding on the vital force (generally blood) of the living." Vampire bats have only recently become part of traditional vampire lore. They were integrated into existing vampire folklore after their discovery in South America in the 16th century. Along with owls, bats have long been associated with the supernatural because of their nocturnal habits. The literary figure Count Dracula is the title character of Bram Stoker's 1897 horror novel. In the novel, he transforms himself several times into a bat. Vampire bats are mentioned twice in the book.

Finally, let's discuss a few additional facts shared by all bats. Already some 50–60 million years ago, bats had achieved a status of sophisticated design as perfect as in recent species.[251] There are more than 900 different bat species indicating that they enjoyed a flourishing existence for a very long time. Their success was enabled by avoidance of faster flying birds by restricting their foraging flights to the night. They can hunt in the dark because of their advanced echolocation capabilities. There are bat species that echolocate within a wide range of sound frequencies, adapted to their environment and prey types. They range from as low as 11 kHz to as high as 212 kHz. Vampire bats will use low frequencies, while insectivorous aerial-hawking bats have a call frequency between 20 kHz and 60 kHz because it is the frequency that gives the best range and image acuity and makes them less conspicuous to insects. In the bat–moth (predator–prey) arms race, some insects have evolved countermeasures, either trying to dampen the sound waves[252] used by bat echolocation to escape detection, or by diverting bat attack via reflecting sonar by spinning hindwing tails (silk moth) in order to create a misleading echoic target.[253]

The intensity of echolocation calls can vary between 60 and 140 dB. Microbat species can modify their call intensity mid-call, lowering the intensity as they approach objects that reflect sound strongly. This prevents the returning echo from deafening the bat. High-intensity calls such as those from aerial-hawking bats (133 dB) are adaptive to hunting in open skies.

Electric Eel (Electrophorus electricus) *and Ghost Knifefish (Apteronotidae)*

The electric eel (Fig. 6C.12a) is a South American "electric fish." Despite the name, it is not an eel, but rather a knifefish.

Figure 6C.12a Electric eel, a fish that both tracks and attacks its prey with electric shocks (source: Wikipedia).

When originally described by Carl Linnaeus in 1766, he used the name *Gymnotus electricus*, placing it in the same genus as *Gymnotus carapo* (banded knifefish) which he had described several years earlier. A century later, in 1864, the electric eel was moved to its own genus, Electrophorus. Later, however, the electric eel was again merged back into the family Gymnotidae, alongside Gymnotus.

The electric eels are rather large, typically growing to about 2 m in length and 20 kg in weight. Their coloration is dark gray-brown on the back and yellow or orange on the belly. They have no scales. *E. electricus* has a well-developed sense of hearing, but does not see well. The respiratory system is very interesting: electric eels are "obligate air-breathers," i.e., they must rise to the surface about every 10 minutes to inhale before returning to the bottom. Nearly 80% of their oxygen needs are covered this way.[254]

What makes the electric eel so special is its "laser-like" electric capability. *E. electricus* has three pairs of abdominal organs that produce electricity: the main organ, Hunter's organ, and Sach's organ. The main and Hunter's organs are the high-voltage producers, used for protection, fright reflexes, and stunning prey, while Sach's organ is capable of producing low-voltage pulses that are used for electrocommunication and navigation. Together, these organs make up 80% of its body and give the electric eel the ability to generate electric discharges of various types. These electric organs are made of electrocytes, modified muscle or nerve cells that generate electricity. They are lined up or "stacked" so a current of ions can flow through them. Each layer adds to a potential difference.[255] In the electric eel's high voltage organs, some 5000-6000 stacked electroplaques can generate an electric shock of up to 860 V and 1 A of current (860 W) for 2 ms.[256]

When the eel finds its prey, the brain sends a signal through the nervous system to the electrocytes. This opens the ion channels, allowing sodium to flow through, reversing the polarity momentarily. By causing a sudden difference in electric potential, it generates an electric current in a manner similar to a battery, in which stacked plates produce an electric potential difference.

The low-voltage (~10 V) Sach's organ of *E. electricus* is associated with electrolocation. For this purpose the electrical signal is discharged at a frequency around 25 Hz.

Hunter's organ can emit high-voltage signals at rates of several hundred hertz.

According to Ken Catania of Vanderbilt University, Tennessee, the electric eel relies on its strong electrical bursts to both track and attack its prey.[257] The strong pulses are both a weapon and a sensory system: first the electric eel emits two pulses to make its prey twitch (and reveal its location); then it releases hundreds of strong pulses to freeze the prey's muscles and to help the eel track the prey's exact location.

Researchers at Yale University and the National Institute of Standards and Technology (NIST) are working on artificial cells that could replicate and perhaps even improve the electrical properties of electric eel cells. Such artificial versions of the eel's electricity-generating cells could power medical implants and other microscopic devices.[258]

The **ghost knifefish** is another electric relative of the electric eel.

It may not achieve the record high voltages generated by *E. electricus*, but it is able to discharge electricity at frequencies approaching 2000 Hz, the most in the animal kingdom.

Ghost knifefishes (Apteronotidae) (Fig. 6C.12b) are a family of ray-finned fishes in the order Gymnotiformes. They can be found in the freshwater of Panama and South America and use a high-frequency wave-type electric organ discharge (EOD) to communicate.

Figure 6C.12b The ghost knifefish discharges electricity at 2000 Hz (source: Wikipedia).

To find out how the ghost knifefish does this, researchers[259] compared the gene encoding the voltage-gated sodium channel, a protein essential for generating electrical signals, in the ghost knifefish with those of the glass knifefish, an electric relative, and the channel catfish, a nonelectric species. They concluded that this gene was first duplicated in ghost knifefish approximately 14.5 million years ago, and then underwent several mutations that hade the following effects:

- made the voltage-gated sodium channel protein fire more frequently
- recruited both nerve cells in the spinal cord and muscle cells to synthesize the protein

The duplicated gene copy gained expression in the spinal cord, where the electric organ is located.

The study[260] shows that duplicate genes can quickly adapt to contribute to the evolution of novel organ systems.

The ghost knifefish uses these electrical sparks to navigate, detect objects, and communicate.

In addition to Robotics research based on the undulating fins of ghost knifefish, these findings could provide new clues about the genetic basis of epilepsy and certain inherited muscle diseases, which are associated with mutations in sodium channel genes.

The Dolphin (Cetacea/Delphinidae, etc.)

The family of dolphins (Fig. 6C.13) includes the Delphinidae (the oceanic dolphins), Platanistidae (the Indian river dolphins), Iniidae (the new world river dolphins), Pontoporiidae (the brackish dolphins), and the extinct Lipotidae (baiji or Chinese river dolphin). More about the Baiji further below. Dolphins are cetaceans, and cetaceans' closest living relatives are the hippopotamuses, having diverged about 40 million years ago. In casual language, the name "dolphin" is normally used as a synonym for bottlenose dolphin, the most common species of dolphin.

Figure 6C.13 Jumping bottlenose dolphin, the most familiar of the Cetacea (source: Wikipedia).

Dolphins range in size from the 1.7 m long and 50 kg Maui's dolphin to the 9.5 m long and 10,000 kg killer whale. They have streamlined bodies and two limbs that are modified into flippers. Though not quite as flexible as seals, dolphins travel at over 30 km/h. The killer whale (Orca) and Dall's purpoise can swim as fast as 55 km/h.

They have well-developed hearing which is adapted for both air and water and is functioning so well that some can survive even if they are blind. Some species are able to dive to great depths and will slow their heart rate to conserve oxygen. Some can also re-route blood from tissue tolerant of water pressure to the heart, brain, and other organs. They are also storing hemoglobin and myoglobin in body tissues to prolong the time they can spend under water. Dolphins feed largely on fish and squid, but a few, like the killer whale, feed on large mammals, like seals, and sometimes even on dolphins. Male dolphins typically mate with multiple females every year, but females only mate every two to three years. Mothers of some species fast and nurse their young for a relatively long period of time. Dolphins produce a variety of vocalizations, usually in the form of clicks and whistles.

Their eyes are relatively small, but they still have good eyesight. Their vision consists of two fields, one on each side, rather than a binocular view like humans and birds. Outside the water, their lens and cornea correct the nearsightedness that results from the changing refraction of light. They have a large number of rod cells to see in dim light, but their cone cells have limited functionality: dolphins lack short wavelength sensitive pigments leading to a rather limited capacity for color vision. The olfactory lobes are absent in dolphins, suggesting that they have no sense of smell.

However, dolphins seem to have a sense of taste although their taste buds are atrophied or missing altogether.

While terrestrial primates rely on vision as the primary sense, for dolphins it's echolocation that dominates their brain's sensory perception. Their dependence on sound processing is evident in the structure of their brain: its neural area devoted to visual imaging is only about one-tenth that of the human brain, while the area devoted to acoustical imaging is about 10 times as large.

Dolphins are capable of making a broad range of sounds using nasal air sacs located just below the blowhole (Fig. 6C.14).

Dolphins send out high frequency clicks from an organ known as a melon. This melon consists of fat. It allows dolphins to produce biosonar for orientation. There are roughly three categories of sounds:

- frequency modulated whistles: produced by vibrating connective tissue, similar to the way human vocal cords function

- burst-pulsed sounds: nature and extent still subject to research
- clicks: they are directional and are used for echolocation. They often occur in a short series called a click train. The click rate increases when approaching an object of interest. Dolphin echolocation clicks are amongst the loudest sounds made by marine animals.[261]

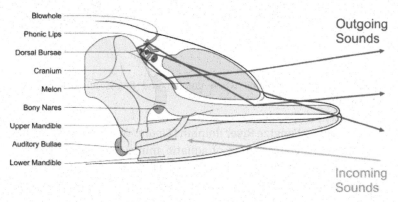

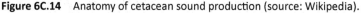

Figure 6C.14   Anatomy of cetacean sound production (source: Wikipedia).

Bottlenose dolphins have been found to have signature whistles, a whistle that is unique to a specific individual and maintained throughout its lifetime. These whistles are used by dolphins for identification, and to communicate with each other. It has even been suggested that dolphins call each other by "name."

Dolphins are very intelligent. They are able to learn, to cooperate, and even to scheme and grieve. The neocortex of many species is home to elongated "spindle neurons" that, in humans, are involved in social conduct, emotions, judgment, and theory of mind.[262] Cetacean spindle neurons are found in areas of the brain that are homologous to where they are found in humans, suggesting that they perform a similar function.

Before discussing topics of potential relevance for human health, let's tell the sad story of the extinct baiji white dolphin, shown in Fig. 6C.15.

Figure 6C.15 (Extinct) baiji white dolphin and finless purpoise (source: Wikipedia).

The baiji, or Yangtze River dolphin (*Lipotes vexillifer*), is a flagship species for the conservation of aquatic animals and ecosystems in the Yangtze River of China. Also known as "Goddess of the Yangtze" or "panda in water," the baiji has become one of the most famous species in aquatic conservation. Unfortunately, this species has experienced a catastrophic population collapse in recent decades due largely to various extreme anthropogenic pressures. Although great efforts have been made to conserve the baiji, the most recent internationally organized survey, which was conducted in late 2006, failed to identify any living individuals. This led the survey organizers to declare it functionally extinct.

Recent surveys indicated that the population size of Yangtze finless porpoise is around 1000 individuals, and it is experiencing an annual population decline of ~13.7%. It has, therefore, since 2013 been considered critically endangered by the IUCN12 and may eventually go extinct, unless effective measures of conservation are urgently implemented.

In 2002, it was estimated there were less than 50 baiji dolphins in the world and it was one of 12 most endangered animals in the world. The situation is more serious than the giant panda. Chances of seeing a Yangtze River dolphin are extremely slim.

The baiji white dolphin had many adaptations. One interesting one was camouflage: it would blend in with the water, uses echolocation to avoid things due to their bad eyesight, they used clicks and whistles like a normal dolphin to communicate. Fossil records suggest that the dolphin first appeared 25 million years ago and migrated from the Pacific Ocean to the Yangtze River 20 million years ago.

Using genomics, scientists found the genome size is about 2.8 Gb, and about 22,168 genes have been annotated. The genomic analysis not only revealed the evolution of the dolphin from terrestrial to aquatic, but also revealed the phylogeny of the baiji. We found a generally slow substitution rate in cetaceans through genome sequencing of the Yangtze River dolphin and comparative analyses with other mammalian genomes. New insights into potential molecular adaptations to secondary aquatic life were outlined, such as a decrease in olfactory and taste receptor genes, changes in vision and hearing genes, and positive selection on genes related to expansion of the brain. Notably, baiji genomes have the lowest density of Single Nucleotide Polymorphisms (SNPs) among the available mammal genomes, which is consistent with the small and rapidly declining population size of this organism. The reconstructed demographic history over the last 100,000 years featured a continual population contraction through the last glacial maximum, a serious bottleneck during the last deglaciation, and sustained and rapid population growth after eustatic sea levels (reflecting a change in the quantity of water in the oceans) approached the current levels. The close correlation between population trends, regional temperatures, and eustatic sea level suggests a dominant role for global and local climate changes in shaping the baiji's ancient population demography. Future genetic sequencing from additional individuals would allow for further reconstruction of the demographic history of the baiji, allow us to test our hypothesis regarding the correlation between baiji population dynamics and global climate fluctuations, and uncover the role of genetic factors in the functional extinction of this organism.

Dolphins are often considered the second- or third-most intelligent animals after sapiens. Research has yielded lots of information about their life, social structure, and development. Their sense of echolocation has fascinated scientists and engineers, who do not hesitate to study it and to apply the same mechanism to technologies that facilitate human life.

There are many things that we could learn from dolphins, especially these two: electrosensing and amazing "powers of healing." The animals can do much more than echolocate: they can sense the electric fields of other animals! *Electrosensing* is mostly used to detect prey, although it might also be useful for evading attacks. Electrosensing could help dolphins find prey at short distances, such as buried in sand, where echolocation wouldn't be of much use. Inspired by this and based on the information learned from dolphins, humans develop positioning technologies, applied to several kinds of fields, especially military.

Finally, dolphins possess amazing *regenerative* capabilities, which could inspire new anti-bacterial treatments and wound healing for humans. For instance, dolphins can heal quickly and painlessly from shark bites, and they use stem cells to rebuild missing tissue. If a shark were to bite a chunk out of a dolphin, it would not hemorrhage or get infected, and apparently the dolphin would not even feel any pain. This information is of considerable interest to stem cell research scientists in San Diego, California, and elsewhere.[263]

Hummingbirds (Trochilidae)

Hummingbirds (Fig. 6C.15) may be small, but they are exceptionally beautiful and capable of unmatched flying abilities, and therefore often called "jewels in flight."[264] Until Columbus, only the native peoples of the Americas ever were able to watch these tiny, long-beaked birds displaying their amazing flying tricks as they flitted from flower to flower in search of nectar. All the more than 320 species of hummingbirds are found in the Americas and most are confined to the American tropics. Nearly half have been observed in Mexico, thereof 12 living in Mexico all year round. The others are merely passing through on their long yearly migrations that can take them all the way North to Alaska or South to the Andes.

The hummingbird is the smallest known bird in the world and the only bird that can fly forward, backward, or upside down, and hover in the air. The name comes from the fact that "hummers" flap their wings so fast (about 80 times per second) that they make a humming noise. The hummingbird's shoulder joints are very flexible, and its wings can go unusually high and low, vibrating forward and backward, and spinning as far as possible, compared with other birds' wings. Their "hovering" is achieved by rapidly alternating forward and backward strokes of their wings. No other bird is capable of their full repertoire of tricks. As much as 30% of the hummingbird's body weight is in muscles devoted to sustaining flight. They also have the most rapid heartbeat of all birds, 500 beats per minute at rest and up to 1260 beats per minute in flight. Here is another top performance: Relative to their size, they are the fastest birds in the world. They can fly up to 55 km per hour and, during courtship, *dive* at speeds of about 100 km/h, pulling more *g*'s[j] (up to 10) than a fighter pilot. In fact, a human fighter pilot would probably become unconscious (pass out) under similar conditions. The peregrine falcon may be faster in absolute terms, but the hummingbird beats all flying objects relative to its small size.

Figure 6C.16 Hummingbirds are "jewels in flight" (source: Wikipedia).

The hummingbirds' advanced way of flying or hovering also requires visual capabilities not shared by ordinary birds.

---

[j]One *g* is the acceleration due to gravity at the Earth's surface and is the standard gravity defined as 9.80665 meters per second squared, or equivalently 9.80665 newtons of force per kilogram of mass.

They have adapted to their navigational needs by development of an exceptionally dense array of retinal neurons. Therefore, hummingbirds have increased spatial resolution in their lateral and frontal visual fields.[265] They have developed a neuronal hypertrophy, relatively the largest in any bird, that is responsible for their refined *dynamic* visual processing while hovering and during rapid flight. Given the fact that only high-speed cameras can capture the details of hummingbird flight, it is understandable that hummingbirds have developed visual perception capabilities not shared by any other bird. Their enhanced visual processing of fast-moving visual stimuli have led to the enlargement of the hummingbird brain region responsible for visual perception. It enables them to conduct rapid forward flight, insect foraging, competitive interactions, and high-speed courtship. Their visual capabilities also enable them to avoid collisions.

Hummingbirds metabolize fastest of all animals. They need to because they consume between one half and up to one and a half times their body weight in nectar each day. Adult hummingbirds live almost entirely on the glucose obtained from nectar. The advantage of nectar foods is that glucose enters the blood immediately to participate in metabolism, but the disadvantage is that the energy is consumed quickly. Fortunately, their raging metabolism slows during the hours of darkness so that they can rest and when necessary they can go into a state of stasis for short periods.

For hummingbirds, there's a direct correlation between the intensity of their energy burn and their need for an energy return. In order to get the huge amounts of food they require, they must feed on hundreds of flowers every day. Their beaks have evolved and are shaped such that they can reach the nectar of those flowers, even when it cannot be reached by any other bird or insect. The hummingbirds have developed a unique symbiotic relationship with plants.

Hummingbirds do not spend all day flying, as the energy cost would be prohibitive; the majority of their activity consists simply of sitting or perching. Hummingbirds eat many small meals. They spend an average of 20–25% of their time flying and feeding, and 75–80% sitting and digesting.

Because their high metabolism makes them vulnerable to starvation, hummingbirds are highly focused on their food sources.

Some species are territorial and will try to guard food sources (such as a feeder) against other hummingbirds, attempting to ensure a future food supply for itself. Compared to other birds, their hippocampus area is two to five times larger.[266] As we have seen before, the hippocampus is a brain region that (across all species) is associated with long-term and spatial memory. Hummingbirds use it to map flowers previously visited during nectar foraging. Although the brains of hummingbirds are as small as only a grain of rice, their memory ability is outstanding: hummingbirds can remember not only the kind of food they've just eaten, but also when they did eat it. These amazing facts seem to make hummingbirds the only wild animal that can remember where and when to have eaten. Previously, scientists believed that only humans could have similar judgment.

The hummingbird's hollow bones, fused spines, and fused pelvis eliminate excessive muscle and ligaments and reduce the bird's weight without sacrificing support for protecting its internal organs. The small foot reduces aerodynamic drag and weight in flight. The longer, stronger bones of the wing keep it stable at each stroke and allow greater fine-grained movement to control flight direction. A more effective pump that enlarges the heart supports faster wing beats and more efficient oxygen distribution to the muscles.

How can a hummingbird weighing less than a tenth of an egg ensure that it doesn't run out of power as it flies? It turns out that there is more to its efficient metabolism than high sugar nectar, supplemented by insect predators to add protein and other nutrients. During the non-breeding daytime, hummingbirds basically only visit flowers to absorb honey and squat to digest. A few minutes later the nectar in the stomach can burn energy in the muscles. However, in the evening they use the sugar they eat for fat synthesis and charge for the next day. After a night of hunger, early morning hummingbirds rely on stored fat metabolism capacity. Also, before a long migration, hummingbirds may have stored 40 percent of their body weight as fat: that's about two grams and all the fuel they need for their long-distance nonstop flights.[267] The small size of hummingbirds may actually be an advantage in long haul flights that can exceed 2000 km. Being small means that they are able to carry relatively more fat than larger birds.

Hummingbirds further have a tricky way of drinking while hovering near a flower. Their tongues have tubes which run down their lengths and help the hummingbirds drink the nectar. High-speed photography has revealed that the tubes open down their sides as the tongue goes into the nectar, and then close around the nectar, trapping it so it can be pulled back into the beak. The forked tongue is compressed until it reaches nectar, then the tongue springs open, the rapid action traps the nectar and the nectar moves up the grooves, like a pump action. There seems to be no capillary action involved.

Strangely, these small, innocent and inoffensive creatures were frequently associated by the early civilizations of Central America with war, blood and human sacrifice. The gentle Toltec god Quetzalcoatl is often pictured with hummingbirds, and the name of the terrible Aztec war god, Huitzilopochtli, actually means "hummingbird on the left."[268] There was even a temple in Tenochtitlan (today's Mexico City) dedicated to the cut-off hummingbird's head, and it was believed that the souls of warriors who died in battle turned into hummingbirds when their task of transporting the sun was finished. The instruments for ritual bloodletting were often in the shape of hummingbirds with needle-sharp beaks. It was also believed that an amulet of dried, stuffed hummingbirds could somehow increase the wearer's sexual performance. Mayan paintings, however, commonly showed the nectar seeking birds with a flower midway down their long beaks.

Hummingbirds in the US have the benefit of being garden and backyard favorites.[269] Many people put out hummingbird feeders that attract hummingbirds in the warmer months to allow these birds to refuel during their long migratory journeys. Many fans of hummingbirds are doing what they can to keep parks and gardens a friendly place for these beautiful birds.

Finally, let's see how far humans have come in trying to emulate the amazing wonders of hummingbird flight. The typical multi rotor unmanned aerial vehicle (UAV) toy, commonly known as a drone, can only stay in the air for about 20 minutes. The hummingbird, specifically the ruby-throated hummingbird (Fig. 6C.17) is only 7 cm long but is able to fly 20 hours without interruption across the Gulf of Mexico, 1000 km wide, on 2 g of fat fuel.

Figure 6C.17 Ruby-throated hummingbird in flight, hovering (source: Wikipedia).

The Peregrine Falcon (Falco peregrinus)

The peregrine falcon (Fig. 6C.18) is a bird of prey (raptor) and member of the family Falconidae. The peregrine is quite large, has a blue-grey or brown back, barred white underparts, and a black head. It has a body length of about 35–60 cm and a wingspan from 75–120 cm. Peregrine falcons are sexually dimorphic, with females being considerably larger than males. The peregrine is renowned for its speed, it is the fastest member of the animal kingdom. During high-speed dives, it can reach amazing velocities.[270] The highest measured speed of a peregrine falcon is 389 km/h.[k]

The peregrine can be found nearly everywhere on Earth, except extreme polar regions, very high mountains, and most tropical rainforests. It is the world's most widespread raptor. Its diet consists mostly of medium-sized birds, but it will also hunt small mammals, small reptiles, or even insects.

The peregrine falcon has been used in falconry for more than 3000 years, beginning with nomads in central Asia. The peregrine is the national animal of the United Arab Emirates. It is admired for its

[k]According to National Geographic.

athleticism and eagerness to hunt. It is also quite easy to train and has a natural flight style of circling above the falconer ("waiting on") before performing an effective and exciting high-speed dive to attack the prey with a fist-like clenched talon.

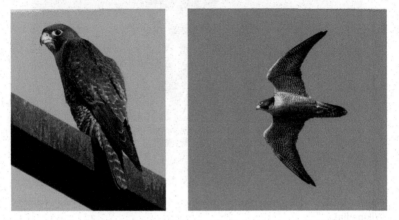

Figure 6C.18 The peregrine falcon is the fastest member of the animal kingdom (source: Wikipedia).

As all birds of prey, peregrines have excellent long-distance "binocular" vision, almost equaling eagles who can see clearly about eight times as far as humans. They are also able to essentially "zoom" in on their prey. Like most birds, they possess a fourth type of color receptor (cone) that enables them to see ultraviolet (UV) light, which is invisible to humans. This may allow them to see urine trails left by some prey animals such as mice. Such urine trails contain elements that reflect UV light and are like "neon signs" that lead the raptors right to their prey. They can also differentiate small changes in coloration in their prey. However, the peregrine's night vision is not as good as the owl's.

To protect their precious and relatively large eyes, the eyes of raptors are surrounded by bone, to both protect the eye and fix it within the skull. Therefore, birds of prey can't move their eyes, must move the entire head to look left and right, up or down. They are equipped with flexible necks and are able to turn their heads a lot more than humans can. In addition, peregrine and other raptor eyes are equipped with three eyelids. Two of their eyelids move up and down to close their eyes, like in humans. The third eyelid is known as a *nictitating membrane,* and this is the eyelid they use to blink with. The nictitating membrane (from Latin *nictare,* to blink) is a transparent or translucent third eyelid present in some animals that can be drawn across the eye for protection and to moisten it while maintaining vision. It is a thin, semitransparent lid, which moves from side to side. When closed, it also helps to protect the bird's eyes but still allows some vision. Raptors close their nictitating membrane before they approach their prey, when in a fast dive, during heavy rain, and even when feeding their chicks. When peregrine falcons go into their high-speed dives, they will blink repeatedly with their nictitating membranes to spread moisture across the eyes and to clear the eye from debris.

Finally, birds of prey have a bony ridge above the eye. Humans also have a bony ridge, but it is less pronounced. In raptors, this helps to protect them from foreign objects that may poke their eyes, such as branches. This bony ridge gives raptors an "angry" or menacing look but also acts like a built-in sunshade.

In conclusion, the peregrine falcon has what it takes (speed and eyesight) to be a successful hunter. In addition, its respiratory system is adapted to its record-breaking diving speed: the peregrine falcon maintains a one-way flow of air so that it can breathe while flying.[271] It also has cones in its nostrils to help regulate breathing at high speeds. Small bony tubercles guide the air entering the nostrils and enable the bird to breathe easily while diving by reducing the change in air pressure

Human engineers of jet engines have actually learned from the peregrine how to improve jet engine airflow: by fashioning a "peregrine-like" cone in the opening of the jet engine, they discovered that the air could pass into the engine even at great speed. Animal adaptation preceded a human invention!

The Bar-Headed Goose (Anser indicus)

The bar-headed goose (Fig. 6C.19) is a goose that breeds in Central Asia in colonies of thousands near mountain lakes. It spends winters in South Asia, as far south as peninsular India. This bird is known to fly across the Himalayan mountains,[272] overcoming the dual challenges of high altitude and cold temperatures.

Figure 6C.19 Bar-headed goose, one of the highest-flying birds known to cross the Himalayas (source: Wikipedia).

The bar-headed goose is pale grey and is easily recognized by the black bars on its head. In flight, its call is a typical goose honking. A mid-sized goose, it measures about 75 cm in total length and weighs up to about 3 kg. In summer, it can be found near high-altitude lakes in Tibet, Kazakhstan, Mongolia, and Russia, where it grazes on short grass.

While some stories of climbers having seen bar-headed geese flying above 8000 m may be exaggerated,[273] it is undeniable that they are able to fly routinely across the world's highest mountain range. Several scientific projects have tried to answer the question how they are able to do it. In a recent paper[274] authored by Scott et al., the Canadian-UK-Australian team concludes that their ability depends on the unique cardiorespiratory physiology of birds in general along with several evolved specializations across the O_2 transport cascade.

Bar-headed geese face several challenges when crossing at altitudes above 6,000 m and sometimes exceeding 7,000 m, namely

- sustaining the high rates of O_2 consumption needed for flapping flight, which ranges from 10- to 15-fold above resting levels during steady flight in a wind tunnel at sea level
- overcoming hypoxic conditions with O_2 levels reduced to one half or less, as compared to sea level
- withstanding temperatures at high altitudes that can be well below freezing year round in the high Himalayas, requiring additional metabolic energy for thermogenesis if the heat

production from exercise is not sufficient to maintain body temperature

• maintaining water balance during flight in the dry air at high altitudes
• dealing with reduced lift when flapping wings in thin air

As bar-headed geese fly to higher elevations, it becomes progressively more difficult to generate lift in the decreasing air density. The geese will have to spend a greater proportion of time flying with near maximal heart rates (above 350 beats per minute) when the altitude exceeds 4800 m. When possible, geese will try to take lower-altitude routes, such as river valleys, or they will take advantage of winds near mountains.[275] Still, bar-headed geese are flapping fliers that very rarely glide, even during steep descent.

Bar-headed geese often migrate at night and in the early morning when the predominant winds travel downslope. This seems counterintuitive because nighttime flights entail a greater metabolic cost than flying later in the day when updrafts predominate. However, the darkness should lessen predation risk, the wind currents are more stable and less turbulent, and the cooler air will have a slightly higher density.

The ability of bar-headed geese to transport enough O_2 to their muscles and other tissues to sustain the high metabolic requirements of flight appears to require *additional traits* that go beyond general traits common to all birds, and have evolved during the process of evolutionary adaptation to high altitude.

The unique physiology of the pulmonary vessels in birds imparts resistance to high-altitude pulmonary edema, a major contributor to acute mountain sickness in mammals. In mammals, the pulmonary vessels constrict in response to hypoxia. This can result in pulmonary hypertension, impairment of gas exchange, and eventually pulmonary edema.

In birds, however, the pulmonary vasculature does not constrict and pulmonary arterial pressures increase in hypoxia only when cardiac output rises. The avian blood-gas barrier is also thought to be mechanically stronger and more resistant to stress failure than that of mammals.[276]

There are also several differences in the hearts of birds compared with mammals that should help support higher cardiac outputs and increased O_2 delivery during hypoxia. Birds have ~50% larger hearts and cardiac stroke volumes than mammals of similar body size, and birds can sustain heart rates during free flight that are similar to or greater than those of mammals during maximal exercise. Capillary density in the cardiac muscle also appears to be higher in birds compared with mammals. As a consequence, birds have a higher oxygen diffusion capacity and the hearts of birds are more resistant to cardiac oxygen limitation in hypoxia.

The capacity for oxygen diffusion into the peripheral tissues also appears to be higher in birds than in mammals. This distinction exists largely because there is a mesh of branching capillaries that surrounds avian muscle fibers, which are themselves smaller in size compared with nonflying mammals of a similar body size.

Another mammalian high altitude challenge is associated with cerebral dysfunction. Again, birds have an advantage: several differences in the brain physiology of birds compared with mammals may protect against cerebral dysfunction under hypoxic conditions, at high altitude. In birds, unlike in mammals, cerebral blood flow is not inhibited by respiratory hypocapnia, defined as a state of reduced carbon dioxide in the blood.[1] This should improve brain oxygenation during environmental hypoxia. Avian neurons also appear to have an inherently higher tolerance of low cellular oxygen levels, and therefore appear to be well protected from cellular damage induced by O_2 limitation. A still unanswered question is whether birds can suffer hypoxic cerebral edema, one of the biggest risks of human high-altitude exposure.

So far, we have discussed birds in general, not yet the additional adaptations that are specific to the bar-headed goose. Bar-headed geese are distinguished by their capacity to transport and consume oxygen at high rates in hypoxia. They can tolerate extreme hypoxia at rest, far exceeding the tolerance of many lowland waterfowl. They also maintain body temperature in hypoxia, and they depress body temperature less. In fact, they elevate their metabolic rate two- to threefold in hypoxia at rest, to support the oxygen demands of the respiratory and cardiovascular responses to hypoxia. Bar-headed geese are also capable of achieving the high metabolic rates needed for flight in a normobaric wind-tunnel at comparable levels of hypoxia to those on the summit of Mount Everest (~7 kPa). As

---

[1]Hypocapnia usually results from deep or rapid breathing, known as hyperventilation.

discussed below, this impressive ability to sustain high metabolic rates in hypoxia appears to arise from increases in the capacity of several steps in the O_2 transport pathway.

Here is a list of known adaptations:

- The control of breathing has evolved in bar-headed geese to improve oxygen uptake into the respiratory system in hypoxia. They exhibit larger increases in total ventilation in response to severe hypoxia than any other bird species studied to date.
- Bar-headed geese breathe more deeply (with higher tidal volumes) and less frequently than low-altitude birds at a given level of total ventilation.
- Bar-headed geese also have ~25% larger lungs than lowland waterfowl of comparable body mass, which should enhance the area and diffusion capacity of the pulmonary gas-exchange surface.
- Circulatory O_2 delivery in hypoxia is improved in bar-headed geese by evolved changes in blood physiology. The hemoglobin of bar-headed geese has a higher affinity for oxygen than that of closely related lowland geese which increases pulmonary oxygen loading and peripheral O_2 delivery in hypoxia by increasing hemoglobin saturation.
- The genetic basis for this increase in affinity could involve several amino-acid substitutions in the α-subunit of the hemoglobin protein. Birds possess major (HbA) and minor (HbD) forms of hemoglobin, and in bar-headed geese the α-subunits of these forms contain four (αA) and two (αD) derived substitutions, respectively. Hemoglobin-O_2 binding is also more sensitive to temperature in bar-headed geese than in other birds and mammals.
- The left-ventricle of the heart, which is responsible for pumping oxygenated blood to the body via systemic circulation, has significantly more capillaries in bar-headed geese compared with lowland birds.
- Compared to lowland birds, mitochondria (the main site of oxygen consumption) in the flight muscle of bar-headed geese are significantly closer to the sarcolemma,[m] decreasing the

---

[m]The sarcolemma, also called the myolemma, is the cell membrane of a striated muscle fiber cell.

intracellular diffusion distance of oxygen from the capillaries to the mitochondria.

- Bar-headed geese have a slightly larger wing area for their weight than other geese, which is also believed to help them generate lift at high altitudes.

In conclusion, we have learned a lot but still want to answer the following questions:

- How are bar-headed geese dealing with low barometric pressure, cold, and dry air at high altitudes?
- What about phenotypic plasticity (i.e., acclimatization) which seems to be so very important for humans climbing 8000 m peaks?

We therefore have much yet to learn about the migration of this fascinating species, crowding mountain lakes in summer (Fig. 6C.20), which will undoubtedly continue to shed light on nature's impressive solutions to oxygen deprivation. It may perhaps even help humans to overcome our existing hypoxia challenges.

Figure 6C.20 Bar-headed geese gathering in big numbers (source: Wikipedia).

The Wild Yak (Bos mutus) *and Domestic Yak* (Bos grunniens)

The wild yak (Fig. 6C.21) is s a large wild bovid[n] that is native to the Himalayan mountains.

---

[n]The **Bovidae** are cloven-hoofed, ruminant mammals that include bison, African buffalo, water buffalo, antelopes, wildebeest, impala, gazelles, sheep, goats, muskoxen, and domestic cattle.

Figure 6C.21 Wild yak, ancestor of the domestic yak (Himalaya) (source: Wikipedia).

The wild yak is the ancestor of the domestic Yak. Wild Yaks are also the largest native animals in their range. Adult wild yaks are formidable animals, up to 2.2 m tall at the shoulder, up to 3.3 m long, and weighing up to 1000 kg. Yak females are about 30% smaller in their linear dimensions and weigh only one third compared to wild Yak bulls. All wild yaks have long shaggy hair with a woolly undercoat for insulation against the cold. This undercoat may form a long "skirt" that can reach the ground. Wild yaks are typically black or dark brown, but some are golden-brown. Wild yaks with golden hair are known as the "wild golden yak." The primary habitat of wild yaks consists of treeless uplands (tundras) in Tibet, Xinjiang, and Ladakh at altitudes between 3000 and 5500 m. They are considered extinct in Nepal and Bhutan.

The domestic yak is descended from the wild yak and can be found throughout the Himalayan region of the Indian subcontinent, the Tibetan Plateau, and as far north as Mongolia and Russia. Domestic yaks can be quite variable in color, including rusty brown, cream, grey, and white (see Fig. 6C.22). They are smaller than their wild relatives: bulls reach weights of up to 600 kg and females weigh up to 250 kg. Yaks are robust cattle with a bulky frame, short but thick legs, and rounded cloven hooves that are splayed to help them walk in snow.[277]

We have included the yak because of its adaptation to a high altitude–low oxygen environment. Their physiology is different from lowland cattle as Yaks have larger lungs and hearts, as well as greater capacity for transporting oxygen through their blood.[278] This appears to be partly due to the persistence of fetal hemoglobin throughout life.[279] Conversely, yaks have problems at lower altitudes and easily suffer from heat exhaustion above about 15°C, partly because they lack functional sweat glands. Yaks only consume the equivalent of 1% of their body weight daily and are able to extract nutrients even from low-quality food.

Figure 6C.22 Domestic yaks colored brown and white. Yaks are sometimes rented to tourists to ride on (source: Wikipedia/Photo by Michael H.).

There are about 12 million domestic yaks in the world, but only 10,000–15,000 wild yaks. On the Tibetan Plateau, the yak is a really unique animal that humans use in many different ways. It is called the "boat of the plateau."

Yaks have relatively small ears, and a broad forehead. Both males and females have horns; in males, the horns sweep out from the sides of the head, and then curve forward, whereas the horns of females are smaller and more upright in shape. The tail is long and similar to a horse's tail, rather than the tail of cattle or bison, which are tufted. Both sexes have a distinctive hump over their shoulders.

Yaks are gregarious animals, and sometimes form herds of up to 200 animals. Most herds are, however, much smaller, including about 10–20 individuals. These herds often only comprise the females and their young, though adult males may sometimes travel with the herd. Most often, however, the males are solitary, or form small bachelor herds. When the conditions are cold, for example at night and in snowstorms, yaks protect themselves from the cold by huddling together, and positioning the calves in the center, where it is warmer. When there is snow on the ground, yaks use their horns to uncover the plants below.

For thousands of years, yaks have provided food (flesh, milk) and animal power for transportation and farming to the native populations of Tibet and other high-altitude areas. After a long period of natural selection, yaks have been able to adapt to high altitude while retaining a considerable degree of genetic diversity. The study of the Yak is important for humans' adaptation to high altitude and for the study of hypoxia and for the study of cardiovascular diseases.

Yaks have 30 pairs of chromosomes and the genome size is about 2.66 Gb. When the yak genome was aligned with the bovine, canine, and human genomes, it was found that 13,810 homologous gene families are shared among the four species.

In order to adapt to hypoxic conditions, yaks have expanded the genome in the corresponding region that is highly expressed in hypoxia. For example, there are 13 copies of a functional domain of HIG (Hikaru genki) gene, and only 9 in cattle, while other mammals have much less. It was found that the genes of Yak's hypoxia adaptation and energy metabolism evolved quite rapidly. Positive selection analysis revealed that there are 3 yak genes, namely ADAM17, ARG2, and MMP3, that are compatible with a high-altitude and low-oxygen environment. These genes can help the yak to adjust and respond to oxygen deficiency or hypoxia. There are also five positive selection genes (CAMK2B, GCNT3, HSD17B12, WHSC1, and GLUL) that can help calves use the limited food on the plateau to obtain energy and adapt to high-altitude environments. The yak has 596 gene family expansions: the families involved in perception and energy metabolism have expanded.

By studying the yak's positive selection genes related to nutrient metabolism, we learn not only about the genetic features behind the important physiological traits of animals in high-altitude areas, but also receive potential clues about the various (hypoxia-related) high-altitude sickness conditions that affect humans. In addition, a careful study of the yak genome will help increase milk production and meat performance of these important economic animals in high-altitude areas.

Finally, here are a few interesting facts[280] about the yak:

- Yaks live at the highest altitude of all mammals (5500 m) and can easily tolerate temperatures of –40°C.

- Yak droppings are used as fuel, and are often the only fuel available on the high Tibetan Plateau, which has no trees. As shown in Fig. 6C.23, they are also used for roof insulation.
- Yak coats have two layers, an outer layer of waterproof, long hair, and an inner layer of very dense, shorter hair that traps the warm air.
- Domestic yaks do not "moo"; they only grunt. Therefore, their Latin name *Bos grunniens* translates as "grunting ox."

Figure 6C.23 Yak droppings are used to insulate roofs of Tibetan houses (Photo by Michael H.).

The Penguins (Sphenisciformes)

The penguins (Fig. 6C.24) are an avian order comprising 18 species. The origin of the name penguin is disputed, but it probably is linked to the Latin word *pinguis*, which means "fat" or "oil." Support for this etymology can be found in the alternative Germanic and Dutch word for penguin, namely "fat goose." Penguins are flightless aquatic birds widely distributed in the Southern Hemisphere. They have completely lost the capacity for aerial flight but are perfectly adapted to their preferred cold seawater environment. They employ modified flipper-like wings in wing-propelled diving or underwater swimming. To adapt to underwater life, evolution has helped penguins to multiple morphological adaptations. For instance, penguins have developed densely packed, scale-like feathers which are good for waterproof and thermal insulation; their eye lens and visual sensitivity are adapted for the efficiency of underwater predation; to overcome buoyancy force in water, penguins have developed dense bones and stiff wing joints, and reduced their wing musculature.

Figure 6C.24 Chinstrap penguin and a group of emperor penguins (source: Wikipedia).

Many penguins have embraced the hostile environments in Antarctica and sub-Antarctica. They are Antarctica's most common birds. Out of eight species that live in the Antarctic and sub-Antarctic areas, two of them (Adélie and emperor penguins) have made the Antarctic continent their major habitat.

The penguins living in Antarctica are subject to extremely low temperatures, high winds, and profound seasonal changes in the length of daylight. To live in such a harsh environment, penguins have developed a complicated system of thermoregulation in their heads, wings, and legs, as well as an effective management of energy storage to support long-term fasting. Because of their important roles in the Antarctic ecosystem and their sensitive responses to changes in marine and Antarctic climate, penguins are also among the widely studied organisms in climate change research.

Penguins' swimming looks very similar to birds' flight in the air. Within the smooth plumage a layer of air is preserved, ensuring buoyancy. This air layer also helps insulate the birds in cold waters. On land, penguins use their tails and wings to maintain balance for their upright stance. All penguins are colored for camouflage: they have black backs and wings with white fronts. Predators such as killer whales or leopard seals may look up from below and see a white penguin belly that is difficult to separate from the reflective water surface. On the other hand, their dark plumage on their back camouflages them from above.

A typical speed reached by diving penguins is 6–12 km/h, but they are able to at least double it in the case of startled flight.

Whereas small penguins do not dive deep, larger penguins can dive very deep indeed in case of need. The large emperor penguin has been recorded to reach a depth of 565 m for up to 22 minutes!

On land, penguins either waddle on their feet or slide on their bellies across the snow while using their feet to propel and steer, a movement called "tobogganing." They also jump with both feet together if they want to move more quickly or have to cross steep and/or rocky terrain.

Penguins have an average sense of hearing, compared to other birds. Their ability to vocalize and to hear is used by parents and chicks to locate one another in crowded colonies. Their eyes are adapted for underwater vision and are their primary means of locating prey and avoiding predators.[281]

The emperor penguins, the largest species, are able to control blood flow to their extremities, reducing the amount of blood that gets cold, but still keeping the extremities from freezing. In the extreme cold of the Antarctic winter, the females are at sea fishing for food while leaving the males to brave the weather by themselves. They often huddle together to keep warm and rotate positions to make sure that each penguin gets a turn in the center of the heat pack.

The remarkable life histories of penguins have attracted broad interest from scientists as well as the general public. They are a beloved species frequently featured in movies and documentaries[282] (Fig. 6C.25).

Penguins are loved around the world, primarily for their unusual, upright, waddling gait, impressive swimming ability, and (compared to other birds) lack of fear of humans. Their striking black-and-white plumage is often likened to a white tie suit.

Most previous studies of penguins have focused on ecological, physiological, behavioral, or phylogenetic aspects of their biology, whereas the molecular genetic basis of penguin adaptations remains largely unknown. As part of the avian phylogenomics project, BGI sequenced the genomes of two Antarctic dwelling penguins (Adélie and emperor penguins) in order to understand the evolutionary history of penguins as well as the genomic and molecular bases of their adaptations to the Antarctic environment.

Figure 6C.25 *Happy Feet* poster and Tux, the (computer software) LINUX mascot (source: Wikipedia).

When BGI scientists sequenced[283] the two Antarctic penguin species in 2014, they looked for new insights into the molecular basis of adaptation to the extreme environment. For example, population genomics analysis of polar bears revealed positively selected genes associated with cardiomyopathy and vascular disease, implying important reorganization of the cardiovascular system in polar bears to adapt to the Arctic environment.[284] The genome of the Tibetan antelope exhibits signals of positive selection and gene-family expansion in genes associated with energy metabolism and oxygen transmission, suggesting high-altitude adaptation in these genes. Furthermore, the midge genome (*Belgica antarctica*) is the first Antarctic eukaryote genome, and has a very compact architecture which is thought to be constrained by environmental extremes in Antarctica. Given their large populations and long history in Antarctica, the emperor and Adélie penguins are excellent models for the study of animal adaption to harsh environments, and for the impact of climate change on population dynamics.

BGI indeed found evidence of associations between these biological patterns and global climate change. In particular, the contrasting patterns in effective population sizes of the two penguins during the last glacial period provide evidence for some previously proposed hypotheses about how different penguin species responded to climate change in the past. Morphological changes in the epidermis and forelimbs are critical for underwater flight in penguins, so the candidate genes that were discovered in this study are highly valuable for future functional studies. The genes involved in light transduction and lipid metabolism exhibit signals of positive selection or pseudogenization, i.e., conversion of a gene into a segment of DNA that no longer codes for a 'gene product'. This result suggests evolutionary responses to the extreme conditions of light and temperature in Antarctica. The pseudogenization events also show examples of relaxed constraints in the two penguins. There are no shared patterns in the molecular evolution of the two penguin species, but there are distinct patterns such as the genes involved in phototransduction and lipid metabolism. This implies that the diversity of molecular evolution in different penguin species deserves further investigation.

The genomic resources and the results presented here lay the foundation for further genomic and molecular studies of penguins. Other "omics" studies, such as transcriptomics and population genomics, are planned in the near future. Future work may involve more in-depth experiments to investigate the functional roles of target genes. Overall, the two penguin genomes will likely facilitate related research, such as penguin biology, avian evolution, polar biology, and climate changes.

Facts about penguins[285]: Our reference names 69 facts, while we are only listing about 20 below.

- Penguins are one of about 40 species of flightless birds. Other flightless birds include rheas, cassowaries, kiwis, ostriches, and emus. Most flightless birds live in the Southern Hemisphere.
- Of the 17 penguin species, 13 are either threatened or endangered, with some on the brink of extinction.
- Male and female penguins look alike.
- Penguins swallow pebbles and stones as well as their food. The stones may help them to grind up and digest their food. The stones may also add enough extra weight to help penguins dive deeper.
- Penguins do not have teeth. They use their beaks to grab and hold wiggling prey. They have spines on the roof of their beak and on their tongues to help them get a good grip.

- Penguins find all their food in the sea and eat mostly fish and squid. They also eat crustaceans, such as crabs and shrimps. A large penguin can collect up to 30 fish in one dive.

- Penguins spend several hours a day preening or caring for their feathers. If penguins don't keep them well maintained, their feathers would not stay waterproof. For extra protection, penguins spread oil, produced by a special gland near their tail, on their feathers.

- Penguins molt, i.e., lose their feathers, once a year. Molting takes weeks and they always molt on land or ice. Until they grow new waterproof coats, they are unable to go into the water. Penguins lose about half their body weight during this time.

- Penguin nesting areas are called "rookeries" and may contain thousands of pairs of birds. Amazingly, each penguin has a distinct call, which allows individual penguins to find their mates and chicks even in the largest groups.

- Penguins are highly social birds, they usually swim and feed in groups. Some penguin colonies on Antarctica can contain millions of penguins at various times during the year.

- A primary reason penguins can swim so fast is that they have a special "bubble boost." When penguins fluff their feathers, they release bubbles that reduce the density of the water around them. The bubbles act as lubrication that decreases water viscosity, similar to competitive swimsuits.

- Most penguins spend nearly 75% of their lives in the water. Penguins stay underwater for, typically, 10–15 minutes before coming to the surface to breathe. Even though penguins spend much of their lives at sea, they all return to land to lay eggs.

- All penguins live in the Southern Hemisphere, from Antarctica to the warmer waters of the Galapagos Islands near the equator.

- To keep from overheating, penguins pant and ruffle their feathers and hold their wings away from their bodies.

- Different penguins species have different ways of attracting a mate. King penguins, for example, sing long songs with their partners. Gentoo Penguin males give their mates gifts of small pebbles or stones.

- Most penguin species lay two eggs. However, emperor and king penguins, the two largest species, build no nest at all and lay just a single egg. They warm their eggs on their feet and cover it with a flap of skin called a "brood pouch."
- When penguin chicks hatch, they are not waterproof, so they must stay out of the ocean. They depend on their parents to bring them food and to keep them warm until waterproof feathers replace their fluffy down coats.
- The earliest known penguin fossil is the *Waimanu manneringi*, which dates from about 60 million years ago. Its name comes from Maori term for "water bird." They were also flightless birds.
- While some penguins mate for life or until a partner dies, some penguins may mate with new partners while the old ones are still alive and in the same colony. Some researchers have noted that male and female penguins sometimes "cheat" on their partners, even while they are nesting and raising young with another penguin.
- Penguins usually enter and leave the sea in large groups, looking for "safety in numbers." By blending into a crowd, an individual penguin may avoid catching the attention of a predator.
- Penguins can drink salt water! They are able to do so because they have a special "supraorbital gland" that filters salt from the bloodstream.
- Penguins have more feathers than most other birds, averaging approximately 70 feathers per square inch. The emperor penguin has the most of any bird, at around 100 feathers per one square inch. 1 square inch = 6.45 cm^2.
- Penguins lost their ability to fly millions of years ago and became the most aquatic of all birds. As such, they are the fastest-swimming and deepest-diving bird species.

The Polar Bear (Ursus maritimus)

The polar bear (Fig. 6C.26) is a carnivorous bear who lives in a range within the Arctic Circle, encompassing the North Pole, its surrounding seas and land masses. It is a large bear: an adult male (boar) weighs up to 700 kg, while an adult female (sow) is about half that size. The polar bear's ecological niche is defined by adaptation to cold temperatures, for moving across snow, ice and open water, and for hunting seals that make up most of its diet. Although most polar bears are born on land, they spend most of their time on the *sea ice*. Polar bears hunt their preferred food of seals from the edge of sea ice, often living off fat reserves when no sea ice is present. Because of their dependence on the sea ice, polar bears are classified as marine mammals, able to swim long distances in cold water. Such swims were recorded off Alaska and the longest measured distance was about 350 km. It took polar bears up to 10 days to travel such long distances. Some researchers believe that the long distance swimming of polar bears and their cubs represents a new adaptation, that there was probably neither the need nor the opportunity to swim such long distances in the past. The polar bear may also swim underwater for up to three minutes to approach seals on shore or on ice floes.

Figure 6C.26　The hyper-carnivorous polar bear (source: Wikipedia).

Polar bears are easily the most hardy carnivores of our planet.

Due to global climate change, polar bears are threatened by the shrinking of available ice flakes. They are now listed as "easily endangered." There are only about 20,000–25,000 polar bears in the world, and the number is decreasing. It is predicted that the population size will be reduced by 30% in the next 35–50 years.

Courtship and mating take place on the sea ice in April and May, typically in the best seal hunting areas. Males may follow the tracks of a breeding female for 100 km or more, and then engage in intense fighting with other males over mating rights. Partners stay together and mate repeatedly for an entire week; the mating ritual induces ovulation in the female.

After mating, the fertilized egg remains in a suspended state until August or September, while the pregnant female eats enormous amounts of food and gains at least 200 kg, often more than doubling her body weight.

When the ice floes are at their minimum in the fall, the pregnant female digs a "maternity den" consisting of a narrow entrance tunnel leading to 1–3 chambers. Maternity dens are typically in snowdrifts or underground in permafrost. They are situated on land a few kilometers from the coast and typically reused each year. In the den, the pregnant female enters a dormant state similar to hibernation. However, this state does not consist of continuous sleeping and the body temperature does not decrease as it would for a typical mammal in hibernation. The heart rate is reduced about 40% (from about 46 to 27 beats per minute).

Between November and February, the cubs will be born. They are blind and will weigh less than 0.9 kg. On average, each litter has two cubs. The family remains in the den, with the mother maintaining her own fast while nursing her cubs on a fat-rich milk. After a few months, when her cubs weigh about 10–15 kg, the polar bear mother will break open the entrance to the den. Mothers show a lot of affection for their cubs and take care of them (Fig. 6C.27a) for two and a half years.

The polar bear has 37 pairs of chromosomes and the genome size is about 2.4 Gb. A comparative genomic analysis of polar bears and brown bears found that polar bears differ from their close relatives, the brown bears, by less than 500,000 years.

An important finding of the polar bear genome analysis is that their high fat intake does not result in abnormalities of the cardiovascular system. We have learned that polar bears have a strongly positive selection of genes related to the cardiovascular system, with focus on fat metabolism and cholesterol metabolism during their adaptation to cold conditions. Polar bear mothers can have a fat content of up to 30%. This is necessary for newly born pups to acquire sufficient nutrients. These fat metabolism related genes have undergone adaptive evolution. Of the 16 strongest positive selection genes, 9 were related to cardiovascular system function. In order to adapt to the extreme cold in the Arctic, polar bears clearly prefer high-fat prey, especially before preparing to hibernate, although the genomic mechanism of hibernation still remains unclear.

Figure 6C.27a   Polar bear cubs are born helpless and typically nurse for two and a half years (source: Wikipedia).

Polar bears are also "experts" in how to use solar energy. Scientists have discovered that polar bears have a strong reflective ability. Surprisingly, by investigating the "white hair" of the polar bear by means of scanning electron microscopy, it was discovered that the hair of the polar bear is not white but a hollow, transparent small light pipe. BGI scientists and their collaborators have found the genes associated with hair pigmentation, and the results were published in *Cell*.[286] In this article, scientists found the special structure of polar bear white hair and black skin, both very important for the absorption of heat from light. Scientists found that the LYST gene plays a key role in the process of pigment metabolism, and that a mutation can lead to pigment metabolism disorder. The mutation of another gene, A1M1, is known to play a very important role in the formation of human melanoma. Both genes are positively selected and may be related to the lack of pigment in polar bear hair.

The peculiar life of polar bears can give us a lot of inspiration. Their body structure has many similarities with the design principles of solar vacuum tube water heaters. And the polar bear's tendency to rely on the warmth of sunshine could also inspire a human energy-saving and emission-reduction lifestyle.

The Cheetah (Acinonyx jubatus)

The cheetah (Fig. 6C.27b) is a large cat of the subfamily Felinae that lives mostly in Africa. By 2016, the global cheetah population has been estimated at approximately 7100 individuals in the wild, already requiring conservation measures.

Figure 6C.27b The cheetah is the fastest land animal. It can accelerate to 75 km/h in ~2 seconds (source: Wikipedia).

It is the fastest land animal. The cheetah has a slender body, spotted coat, small rounded head, long thin legs, and a long spotted tail. Its light build is in sharp contrast with the robust build of the big cats like tigers and lions. The cheetah measures up to 90 cm at the shoulder and weighs up to 72 kg.

Cheetahs are active mainly during the day, with hunting as their major activity. They prey mainly on antelopes and gazelles. As the cheetah lacks the strength of other felids it is unable to fight its prey to the ground. Instead it must trip or pull its prey off balance when traveling at high speeds. Cheetahs can reach speeds of over 100 km/h in short bursts and accelerate to 75 km/h in only 2 seconds, faster than most supercars! They will finally charge towards their chosen prey and kill it by biting its throat to suffocate it to death.

The Cheetah's large nasal passages ensure fast flow of sufficient air, and the enlarged heart and lungs allow the enrichment of blood with oxygen in a short time. Therefore, cheetahs rapidly regain their stamina after a chase. During a typical chase, their respiratory rate increases from 60 to 150 breaths per minute. While running, in addition to using their claws for good traction, they use their tail as a rudder-like means of steering, enabling sharp turns needed to catch up with escaping antelopes. The extension of the vertebral column can add more than 70 cm to the length of a stride. The cheetah's spine is very flexible, which allows them to extend more during a chase for food. The shoulder blades are not attached to the collar bone, which allows the shoulders to move freely. The hips also pivot to allow the rear legs to stretch out further while running. The cheetah is almost "flying" during a sprint, having all four limbs in the air over 50% of the time. A typical sprint will last about 100 m, but it can run at a speed of 80–110 km/h for 500 m. More anatomical and physiological details explaining cheetah performance can be found in a paper by Hudson et al.[287]

The pronghorn (88.5 km/h) and the springbok (88 km/h are competing with the cheetah for the title of "fastest land animal," yet the cheetah has a greater probability of succeeding in their chase due to its exceptional acceleration. One stride or jump of a galloping cheetah averages 6.7 m. In 2012, an 11-year-old cheetah from the Cincinnati Zoo named Sarah set a world record by running 100 m in 5.95 seconds, reaching a recorded maximum speed of 98 km/h.

Cheetahs have a high concentration of nerve cells, arranged in a band or "visual streak" in the center of the eyes. This arrangement significantly enhances the sharpness of the vision and is most concentrated compared to all other felids. The nasal passages are short and large.

The cheetah shows little aggression toward human beings, and can be easily tamed. Tamed cheetahs have appeared in stories and paintings since antiquity. They have even been used by humans to hunt. Some old paintings are believed to depict domesticated hunting cheetahs. As shown in Fig. 6C.28, in *Bacchus and Ariadne*, an oil painting by Tizian, the chariot of the Greek god Dionysus (Bacchus) is drawn by two cheetahs.

Figure 6C.28 *Bacchus and Ariadne* by Tizian (1523) (National Gallery, London/ source: Wikipedia).

In the Middle East, the cheetah would accompany the nobility to hunt in special seats behind saddles. Cheetahs continued to be associated with royalty and elegance in western Asia until as late as the 19th century.

Before closing the chapter on felines, let's discuss the **domestic cat**. All domestic cats descended from the Near Eastern wildcat and diverged around 8000 BC in the Middle East.

Cats can suffer from more than 200 human-like diseases (such as leukemia, atypical pneumonia, diabetes, retinal diseases, spina bifida, etc.), so cat genome research has important implications for human health and pet welfare.

The cat has 19 pairs of chromosomes and the genome size is about 2.7 Gb. A comparison of mammals such as dogs, chimpanzees, and mice revealed 133,499 CSBs (conserved sequence blocks). An important finding of the genomic analysis of cats was the detection of a previously undiscovered retrovirus. It occurs more than ten times more frequently than the number of feline leukemia viruses in the cat genome.

In conclusion, a few heartfelt comments from an apparent cat lover[288]:

Cats are really smart. They have a different view of the world than dogs (see below), but both make our lives richer. While dogs are seemingly more concerned with human happiness, cats are more concerned with their own.

What follows are a few lessons derived from the observation of cats.

- **Cats are independent:** They may rely on food and water provided by the house they live in, but that's about it. They love companionship—up to a point. Whatever the owner's mood, cats are in charge of their own needs and happiness.
- **Cats demand respect:** You are not supposed to wake them from a nap, take away a toy or—much worse—bring a dog home. Dogs may like to see other dogs but cats don't care; they only need themselves. However, they appreciate good food such as fish or birds. They don't need much attention but they appreciate respect.
- **Cats don't know fear:** No problem climbing to very high places like a top kitchen shelf. They will always find a way to get down. Being paralyzed by fear is for lower animals, not cats!
- **Cats keep moving forward:** They will always land on their feet after a fall and definitely don't whine about falling. They don't blame the fall on someone else. They know that they will always manage, and they never expect that bad things will happen to them. All they think about is moving on to their next adventure.

Cats will teach you that obstacles and setbacks do not define you and that you can be in full control of your needs, emotions and goals. Their message: *"There is an amazing world out there!"*

Finally, the biological causes of "cats are natural enemies of rats" have not yet been confirmed.

The Dog (Canis lupus familiaris)

The dog and modern wolves are no longer closely related to the gray wolves that were first domesticated. This implies that the direct ancestor of the dog is extinct. The dog was the first species to be domesticated and has been selectively bred over millennia for various behaviors, sensory capabilities, and physical attributes. Although some researchers held the opinion that the domestic dogs originated about 23,000 years ago in Europe, recent research by an international team of scientists[289] from China, Singapore, Russia, Finland, Sweden, and the United States, has confirmed an ancient origin of domestic dogs in Southern Asia, 33,000 years ago. Using whole genome sequences (instead of just mitochondrial DNA analysis) from a total of 58 canids (12 gray wolves, 27 primitive dogs from Asia and Africa, and a collection of 19 diverse breeds from across the world), the team found that dogs from southern East Asia have significantly higher genetic diversity compared to other populations, and are the most basal group relating to gray wolves. Around 15,000 years ago, a subset of ancestral dogs started migrating to the Middle East, Africa, and Europe, arriving in Europe at about 10,000 years ago. The study unraveled an extraordinary journey that the domestic dog has traveled on Earth.

The dog's wolf ancestors began to associate with people, maybe drawn by food in garbage dumps and carcasses left by human hunters. In the process they became tamer, and scientists believe people found them useful for things like hunting and guard duty. Over a very long time in this human environment, wolves gradually turned into the first dogs. The DNA's of modern dogs show similarities to the genetic material from both ancient and modern-day European wolves. The first dogs evolved by associating with hunter-gatherers rather than farmers, since dogs evidently appeared before agriculture did.[290]

Their long association with humans has led dogs to be uniquely attuned to human behavior, both socially and diet-wise. They are able to thrive on a starch-rich diet that would be inadequate for other wolf-like canid species. Compared to wolves, dogs have smaller brains, and domestication may have reduced their intelligence, while enhancing social-cognitive abilities. Dogs are able to read and react to human body language such as gesturing and pointing, and they understand human voice commands. Dogs also engage in deception, a sign of intelligence. However, wild animals such as wolves and Australian Dingos can outperform domestic dogs in nonsocial problem solving. When dogs are faced with a more difficult version of a known problem, dogs tend to look at the human, while socialized wolves do not. Modern domestic dogs use humans to solve their problems for them. On the other hand, domestic dogs possess social-cognitive abilities that are superior to both the dog's closest canine relatives and to other highly intelligent mammals such as great apes. They are similar to some of the social-cognitive skills of human children. These sophisticated forms of social cognition and communication result in trainability, playfulness, and ability to fit into human households and social situations. Their relationship with humans has enabled them to become one of the most successful species on the planet today.

For humans, the most significant early benefit was to use the dog's robust sense of smell to assist with the hunt. Up to this day, the presence of dogs is a requirement for successful hunting. The cohabitation of dogs and humans did greatly improve the chances of survival for early human groups, and the domestication of dogs may have been one of the key forces that led to human success.

Unfortunately, their close association with humans has also led to increased vulnerability for picking up human diseases such as cancer and diabetes. Dogs vary widely in shape, size, and colors. They perform many roles for people, such as hunting, herding, pulling loads, protection, assisting police and military, companionship and, more recently, aiding handicapped individuals and performing therapeutic roles. No wonder they are called "man's best friend."

Aside from cardiovascular health benefits based on increased exercise, dog owners benefit from their companionship in many other ways, including improved social interactions and mental health. The practice of using dogs and other animals as a part of therapy dates back to the late 18th century, when animals were introduced into mental institutions and nursery homes to help socialize patients with mental disorders. Animal-assisted intervention research has shown that "dog-assisted therapy" can improve social behaviors, such as smiling and laughing, among people with Alzheimer's disease. Also, children with ADHD and conduct disorders who participated in an education program with dogs showed increased attendance and more willingness to reach their knowledge and skill objectives, as well as decreased antisocial and violent behavior.

Modern dog breeds show more variation in size, appearance and behavior than any other domestic animal. Like many other predatory mammals, the dog has powerful muscles and a cardiovascular system that supports both sprinting and endurance. Dogs further have teeth for catching and tearing. The smallest known adult dog was a Yorkshire terrier, that stood only 6.3 cm at the shoulder, 9.5 cm in length along the head and body, and weighed only 113 g.

The largest known dog was an English mastiff weighing 155.6 kg and measuring 250 cm from the snout to the tail. The Great Dane can be even taller, up to more than 100 cm at the shoulder, as shown in Fig. 6C.29.

Figure 6C.29 Dog Breeds vary greatly in size: Yorkshire terriers (left) and a Great Dane (right) (Photo by Michael H./source: Wikipedia).

Here is a list of interesting facts[291] about dogs:

- There are more than 150 dog breeds, divided into 8 classes: sporting, hound, working, terrier, toy, non-sporting, herding, and miscellaneous.
- Dogs have a relatively small number of taste buds (about 1700 on their tongues) compared to humans (we have 9000), but have over 200 million scent receptors in their noses, 40 times more than humans.
- An adult dog has 42 teeth.
- Dogs do see in color, just not as vividly as humans do. Vision is not their greatest asset, their nose is.
- Dogs are capable of locating the source of a sound by using their swiveling ears like radar dishes.

The loyalty to humans that is shown by dogs is legendary. Take for instance the Japanese story of a loyal Akita dog, Hachikō: Hachikō (November 10, 1923–March 8, 1935) was born on a farm near the city of Ōdate, Akita Prefecture, Japan. He is remembered for waiting over nine years following the death of his owner, Prof. Ueno (Fig. 6C.30), at a train station where his master used to arrive in the evening. During his lifetime, the dog was held up in Japanese culture as an example of loyalty and fidelity. Hachikō has been immortalized with a sculpture at the Shibuya station in Tokyo. His story, adapted to a railway station in Rhode Island, USA, has been made into the movie *Hachi*, starring Richard Gere (see rightmost picture of Fig. 6C.30).

Figure 6C.30 Japanese Akita dog Hachikō and his master, Professor Ueno, and a movie poster (source: Wikipedia).

Figure 6C.31 Alaskan and Siberian Husky, the dog of choice for high-speed dog sled racing (source: Wikipedia).

The dog is not only sapiens' most loyal and best friend, but also capable of very high endurance and speed, even in extreme winter conditions while pulling a sled (Fig. 6C.31). Dog sled races have strict rules that have been in effect since the beginning of organized racing in the city of Nome, Alaska, in 1908. The most famous long-distance race is the Iditarod Trail Sled Dog Race, also named the "Last Great

Race on Earth," with roughly 1000 miles of some of very rough but also extremely beautiful terrain. The dogs have to pull sleds across fierce mountains, frozen rivers, thick forests, and desolate tundras. Each team of 12–16 dogs must go from Anchorage to Nome.[292] Similarly, the Balto[293] movies are loosely based on a true story about the dog who helped save children from the diphtheria epidemic in the 1925 serum run to Nome, by carrying medical supplies through difficult terrain in a tough winter storm.

Dog Genetics

To date, a complete genome sequence of a chow and 14 different breeds of domestic dogs and wolves have been published. It shows that the dog genome has 39 pairs of chromosomes and that the genome size is about 2.4 Gb. In evolutionary history, the similarity between dogs and humans is high, and the functional evolution processes of humans and dogs are fairly parallel and similar. Dogs have the lowest insertion rate of transposition factors, and have the lowest base substitution rate compared to humans. The most conserved non-coding sequences in the dog genome are concentrated in more than 200 gene-poor regions.

Mars, a US family business focused on pet food, pet health, and chocolate, is a leader in pet research focused on cats and dogs. In their Waltham, UK, based research facility, they have developed a genetic test[294] to identify dog breeds and their admixtures.

The Platypus (Ornithorhynchus anatinus)

There is a German humoristic expression, "Eierlegendes Woll-Milch-Schwein," that describes a hypothetical animal hybrid hen-sheep-cow-pig that contributes eggs, wool, milk, and pork, i.e., satisfies all kinds of human needs. It is shown in Fig. 6C.32.

When looking at the platypus, it almost feels like evolution tried to create a species that retained features found among birds, reptiles, and mammals: when European naturalists first came to Australia and encountered the unusual appearance of this egg-laying, duck-billed, beaver-tailed, otter-footed mammal, they were baffled. Some even believed they were the victims of a hoax. Anyway, this animal is not a hypothetical hybrid; it is real!

Figure 6C.32   Eierlegendes Woll-Milch-Schwein (source: Wikipedia).

The platypus is the most primitive and peculiar animal among mammals. It is a unique mammalian monotreme° in Australia and one of only three egg producers in the world (the other two are Echidnas, sometimes known as spiny anteaters, such as hedgehogs and acupunctures). The platypus is among nature's most unlikely animals. It is best described as a hodgepodge of more familiar species: the duck (bill and webbed feet), beaver (tail), and otter (body and fur) (Fig. 6C.33). Males are also *venomous!* They have sharp stingers on the heels of their rear feet and can use them to deliver a strong toxic blow to any foe.

The platypus is one of the world's great mysteries. This unique mammal combines the looks and traits of several different animals. Its most pronounced feature is its duck-like bill—hence a common second name for the animal, the "duck-billed platypus." The bill looks like a duck's but is actually much softer and acts almost like

---

°Monotremata, the most primitive order of mammals, characterized by certain birdlike and reptilian features, as hatching young from eggs, and having a single opening for the digestive, urinary, and genital organs, comprising only the duckbill and the echidnas of Australia and New Guinea.

a snout. The platypus uses its snout to scoop up food—usually worms, other insects, and small shrimp—from the bottom of lakes or streams. The platypus also has webbed feet—but the similarities to a duck end there. Its long, muscular tail is similar to a beaver's. Its body and fur look just like an otter's. Males carry venom that can cause considerable pain to humans. And the female platypus lays eggs like a bird or reptile. Also, unlike most mammals, a platypus lacks teeth. The platypus swims for up to 12 hours a day to find food, but otherwise lives on land. Many scientists think they are the earliest-evolving mammal—which might explain their unique characteristics.

Figure 6C.33 Australia's duck-billed platypus, nature's most unlikely animal (source: Wikipedia).

Platypuses hunt underwater, where they swim gracefully by paddling with their front webbed feet and steering with their hind feet and beaver-like tail. Folds of skin cover their eyes and ears to prevent water from entering, and the nostrils close with a watertight seal. In this posture, a platypus can remain submerged for a minute or two and employ its sensitive bill to find food. When swimming, it can be distinguished from other Australian mammals by the absence of visible ears. It propels itself by an alternate rowing motion of the front feet. Although all four feet are webbed, the hind feet (which are held against the body) do not assist in propulsion. They are used for steering in combination with the tail.

These Australian mammals are bottom feeders. They scoop up insects and larvae, shellfish, and worms in their bill along with bits of gravel and mud from the bottom. All this material is stored in cheek pouches and, at the surface, mashed for consumption. The bits of gravel help them to "chew" their meal by grinding it up.[295]

On land, platypuses move a bit more awkwardly. However, the webbing on their feet retracts to expose individual nails and allow the creatures to run. Platypuses use their nails and feet to construct dirt burrows at the water's edge.[296]

Platypus reproduction is quite unique. Females seal themselves inside one of the burrow's chambers to lay their eggs. A mother typically produces one or two eggs and keeps them warm by holding them between her body and her tail. The eggs hatch in about ten days, but platypus infants are the size of lima beans and totally helpless. Females nurse their young for three to four months until the babies can swim on their own.

The platypus genome has 26 pairs of chromosomes and the genome size is about 1.9 Gb. The platypus differed from humans about 166 million years ago. The most significant difference between the two is the gender-determining system. Humans only have 2 sex chromosomes, but the platypus has 10 (5X and 5Y). These gender-determining gene sequences are similar to the bird's Z chromosome. Comparative analysis of genomic sequences have also revealed that the platypus carries a "mixture" of genes, i.e., some bird genes, some reptile genes, and some mammalian genes.

A few other facts[297] about the platypus:

- A platypus finds food by using sensors in its bill to pick up electric impulses from moving sources of food like worms and insects. The electroreceptors are located in rows in the skin of the bill, while mechanoreceptors (which detect touch) are uniformly distributed across the bill.

- Some cortical cells in the brain receive input from both electroreceptors and mechanoreceptors, suggesting a close association between the tactile and electric senses. These receptors dominate the platypus brain in the same way as human hands dominate a part of the human brain.

- The platypus has adapted to an aquatic and nocturnal lifestyle. It has developed its electrosensory system at the expense of its visual system.

- The platypus maintains its body temperature at about 32°C, lower than most mammals, even while spending hours in water below 5°C.

- The platypus is extremely popular in its native Australia, having served as a mascot for athletic events, including the Summer Olympics. It has also been featured on Australian stamps.
- Perhaps the most famous platypus is Perry the Platypus from the popular *Phineas and Ferb* cartoon. Perry lives a secret life as James Bond-like "Agent P."

Lessons learned[298] from the platypus:

Sequencing of the platypus genome reveals that the platypus has only about 18,000 genes. Roughly 82% of the platypus's genes are shared between monotremes, marsupials,[p] eutherians (mammals having a placenta), birds, and reptiles.

Humans and platypuses do differ in many ways. For instance, an obvious difference is that the platypus lays yolky eggs, whereas humans and other eutherians have yolkless eggs that are retained in the mother's body. Thus, as you might expect, the platypus has a gene that humans lack—one that codes for vitellogenin, a crucial yolk protein.

As opposed to the presence of vitellogenin, a trait that both eutherians and monotremes have in common—but one that is not shared with birds—is lactation. In the ancestral state, lactation was probably the secretion of fluids and immune system proteins to keep eggs and newborns hydrated and protected, but parents who invested more effort in secreting additional nutritive components, like sugars, fats, proteins, and calcium, were more successful. Like humans, the platypus secretes a true milk that is loaded with all of these components, including a protein called casein, which is thought to have originated by way of the duplication of a tooth enamel matrix protein gene.

An interesting specialization in the platypus is the evolution of venoms. The platypus has small, sharp spurs on its hind limbs that it uses to inject defensive poisons into predators, an unusual feature not found in other mammals. Where did these venoms come from? As it turns out, they arose through the duplication of genes that have other functions, with subsequent divergence. Many of these genes are involved in the functioning of the platypus's innate immune

---

[p]A mammal of an order whose members are born incompletely developed and are carried and suckled in a pouch on the mother's belly. Marsupials are found mainly in Australia although the opossums live in America.

system. In particular, there is a set of genes in the platypus that code for the production of proteins called b-defensins. The platypus has repurposed the b-defensin genes, making copies that have been selected for more effective toxicity when their product proteins are injected into other animals. These are the same proteins used in venomous reptiles—for instance, snake venoms also contain novel forms of b-defensins. This means that animals from two distantly related groups—the lepidosaurs and the monotremes—both use b-defensin-derived venoms. However, venomous snakes and the platypus have different duplications of the b-defensin genes. So, while co-opting these genes seems to be a common strategy for evolving venoms, the details of the gene duplications reveal that platypus venom and snake venom are independently derived features.

Chapter 7

Current Use and Future Promise of Genetic Engineering

We have already touched on genetic engineering's history when we introduced some basic concepts of molecular and cellular biology in Chapter 2. We learned about Watson and Crick's discovery of the double helix, about DNA transcription and replication, and about the genetic code. Another key advance was the emergence of "recombinant DNA" and gene cloning. New techniques like gene sequencing made it possible to identify genes and to determine their function. What came next was to isolate the segment of DNA that codes for a particular protein, and to then snip it out with molecular scissors—specialized enzymes, and then to make copies of the gene, a process called cloning: the ends of the excised gene are incorporated into stretches of DNA from another organism, preferably one that divides quickly.

The essence of recombinant DNA is the production of large numbers of copies of a gene that has been snipped from one organism and then stitched to the DNA of another organism that serves as the vehicle for replication. It's like cutting and pasting of words in a document.

In 1972, Paul Berg[299] created the first recombinant DNA molecule, and in 1973 Boyer and Cohen[300] developed gene cloning. The first significant application was the splicing of human insulin

Our Animal Connection: What Sapiens Can Learn from Other Species (Second Edition)
Michael Hehenberger and Zhi Xia
Copyright © 2021 Jenny Stanford Publishing Pte. Ltd.
ISBN 978-981-4877-50-3 (Hardcover), 978-1-003-13072-7 (eBook)
www.jennystanford.com

gene into a bacterium, and to then produce unlimited quantities of human insulin by means of harvesting the protein output of the bacterium (*E. coli*).

In the 1970s, the US government declared "war on cancer" and primarily used the National Cancer Institute[301] (NCI) of the National Institutes of Health (NIH) to fund not only specific cancer research, but basic biomedical research. In 1980, both the science and business world learned that the evolving genetic engineering technologies would give rise to the new era of *biotechnology*.

Instead of the traditional pharmaceutical focus on small molecules to cure diseases, biotechnology focused on ways to manufacture proteins and to emulate nature's way to use the genetic code. The tools and methods developed by pioneers such as Genentech and other early biotech companies provided the building blocks for new drugs and treatments. Those techniques were also precursors to the techniques later used to map the human genome.

The biotech industry invented the word "biologic" to describe drugs that were based on proteins and therefore were difficult to administer to patients. However, it turned out that those biologics could often achieve life-saving results impossible to obtain the old-fashioned way. The door was open to translate genomics, proteomics, and cell biology research into benefits for human health. Genetic engineering has enabled amazing new medical breakthroughs. There is no end in sight what we will be able to do. However, new emerging bioethical issues will also have to be confronted.

Before turning to the possible future, let's review where we are today.

Since the beginnings of biotech industry in the 1970s, the primary ways to develop biologics are to use recombinant DNA technology and monoclonal antibodies.

Recombinant DNA molecules are formed by laboratory methods of genetic recombination to bring together genetic material from multiple sources. By doing so, it is possible to create sequences that would not otherwise be found in biological organisms.

The DNA sequences used in the construction of recombinant DNA molecules can originate from a variety of living organisms, including plants, bacteria, fungi, etc. They may include human DNA. It is even possible to create DNA sequences that do not occur anywhere in nature. Using recombinant DNA technology, more or less any DNA sequence may be created and introduced into an extensive range of living organisms.

Recombinant proteins are resulting from the expression of recombinant DNA within living cells. They are often the end result of the recombinant DNA technology.

Recombinant DNA is created by two main methods, polymerase chain reaction (PCR) and cloning. PCR is used to generate thousands to millions of copies of a particular DNA segment. Molecular cloning involves replication of the DNA within a living cell.

Recombinant DNA (rDNA) can also be used to "knock out" genes in the host cell to determine their biological function and importance.

Recombinant DNA is widely used in biotechnology, medicine, and research:

- *Recombinant human insulin* has almost entirely replaced insulin previously obtained from animal sources (e.g., pigs and cattle) for the treatment of diabetes. Recombinant insulin is typically synthesized by inserting the human insulin gene into *E. coli*, which then produces insulin for human use.

- *Recombinant human growth hormone* (HGH, somatotropin) is administered to patients whose pituitary glands generate insufficient quantities. Recombinant HGH must be added to support normal growth and development. Before recombinant HGH became available, HGH for therapeutic use could only be extracted from cadavers. It was an unsafe practice that carried the risk of Creutzfeldt-Jacob disease transfer to patients treated with HGH.

- *Recombinant hepatitis B vaccine* can be used to control hepatitis B infection. The vaccine contains a form of the hepatitis B virus surface antigen that is produced in *yeast cells*. Unlike other common viruses such as polio virus, the hepatitis B virus cannot be grown in vitro. Therefore, the development of the recombinant subunit vaccine was a crucial breakthrough.

- *Epoetin alfa* (Amgen's drug Epogen, FDA approval 1989) is human erythropoietin produced in cell culture using recombinant DNA technology. EPO stimulates erythropoiesis (increases red blood cell levels) and is used to treat anemia, commonly associated with chronic renal failure and cancer chemotherapy.

Recombinant DNA technology[302] sparked the biotech revolution. The first licensed drug generated using recombinant DNA technology was human insulin, developed by Genentech. The product (Humulin) was then licensed to and manufactured by Lilly, and became the first-ever approved *genetically engineered* human therapeutic.

Another important genetic engineering technology that enabled the biotech revolution is DNA Sequencing. *DNA sequencing technologies* are used to determine the order of the nucleotide bases A, G, C, and T in a molecule of DNA. The first DNA sequences were obtained in the early 1970s by academic researchers. Early forms of nucleotide sequencing were based on chromatography, laboratory techniques for the separation of mixtures that were invented in 1900 by Michail Tsvet[303] and first applied by him to the extraction of plant pigments such as chlorophyll and carotene. RNA sequencing was done first because it was easier to deal with a single strand of a helical molecule than working with the full double helix. Between 1972 and 1976, Walter Fiers and his coworkers at Ghent (Belgium) were able to sequence the first complete gene and the complete genome of a viral genome, "Bacteriophage MS2."[304] During the 1970s, Frederick Sanger at the Medical Research Council in Cambridge, UK, developed his method of "DNA sequencing with chain-terminating inhibitors,"[305] and Walter Gilbert and Allan Maxam at Harvard in Cambridge, Massachusetts, developed "DNA sequencing by chemical degradation."[306] In 1980, Sanger and Gilbert shared one half of the Nobel Prize[307] for their "contributions concerning the determination of base sequences in nucleic acids," with Paul Berg of Stanford University capturing the other half for his "fundamental studies of the biochemistry of nucleic acids, with particular regard to recombinant-DNA," as already covered above.

Maxam–Gilbert sequencing is based on chemical modification of DNA and subsequent cleavage at specific bases. Also known as *chemical sequencing*, this method's interesting history is described in detail by Gilbert in his Nobel lecture.[308] The basic idea is to find a "repressor" protein that can protect a short stretch of DNA from degradation, caused, e.g., by digestive enzymes. Subsequently, they found that protein binding would block not only enzymes but also chemical methylation, the addition of CH_3 groups to the bases in DNA. When comparing DNA fragments in the presence and absence of the repressor, there would be a stretch of DNA that would be methylated without repressor but be protected with it. By finding four separate chemical ways to "fracture" DNA selectively at each of the four bases A, G, C, and T, Maxam and Gilbert were able to deduct DNA sequences. However, the method's use of radioactive labeling and its technical complexity discouraged extensive use.

In his Nobel lecture[309] in 1959, Arthur Kornberg stated that "what Sanger has done for protein sequence remains to be done for nucleic acids. The problem is more difficult, but not insoluble." Amazingly, Fredrick Sanger accepted the challenge and succeeded. His method, called the "chain-termination method,"[310] developed by him and a team of coworkers in 1977, soon became the method of choice, owing to its relative ease and reliability.

Sanger's DNA sequencing method was based on his deep understanding of DNA's chemistry: the outer backbone of DNA is a monotonous repeat of sugar-phosphate-sugar-phosphate . . ., the only variables being which base (A, C, T, G) is inserted in the "rungs" of the DNA ladder. Sanger's initial idea was to supply all of the components needed to produce the DNA sequence but to "starve" the reaction of one of the four bases needed to make DNA. For instance, in the sequence ACGTCGGTGC,[311] starving for T would produce ACGTCGG(blank) and ACG(blank). By separating the resulting molecules by length, it could be found that the eighth and fourth position of ACGTCGGTGC should be T.

Likewise, starving for G would produce AC(end), ACGTC(end) and ACGTCGGT(end), meaning G was in position 3, 6, 7, and 9. The whole sequence would follow directly from starving for each of the four nucleotide precursors and separating the molecules by length.

Sanger's breakthrough idea[312] was to find chemicals that were inserted in place of A's, C's, T's and G's, but caused a growing DNA chain to end. For our example, the fragments obtained via "chain termination" would be ACGT* and ACGTCGGT*. With one terminator for each base type, the sequence could be read just by measuring the respective lengths of the fragments.

Sanger sequencing is the method which prevailed from the 1980s until the mid-2000s. Over that period, great technical advances were made and Sanger sequencing was automated. Important contributions were made by Leroy Hood's team at the California Institute of Technology and by the startup company Applied Biosystems (ABI) which was founded in 1981 by two Hewlett-

Packard engineers to build instruments for the new biotechnology. ABI soon became the leading manufacturer of both protein and DNA sequencers, thanks to a very close cooperation with Leroy Hood's molecular biology group at CalTech. The automated DNA sequencing instruments that were commercialized by ABI were still based on Sanger's original concept but included several important modifications, leading to the workflow illustrated by Fig. 7.1:

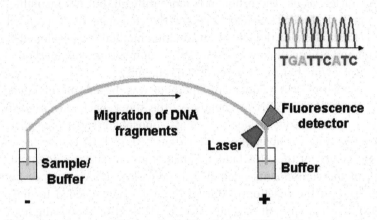

Figure 7.1 Automated Sanger sequencing (source: Wikipedia).

The DNA fragments sequenced by automated Sanger sequencing instruments have to be re-assembled in order to reconstruct the original sequence to be analyzed. The size of Sanger sequencing fragments is limited to about 800–1000 base pairs which has to be compared to a full human genome consisting of 3 billion base pairs. Assembling a full genome is therefore a highly complex task requiring sophisticated computer software and expert bioinformatics skills.

After three decades of gradual improvement, the Sanger biochemistry can achieve per-base "raw" accuracies as high as 99.999%, higher than all "next-generation sequencing" (NGS) technologies, to be covered below. The common NGS goal is to significantly increase sequencing speed and to cut cost with an ultimate objective of achieving the "$1000 ($1K) Human Genome."

Automated Sanger sequencing or "first-generation sequencing" can be characterized as a low-throughput operation with high accuracy, relying on fluorescent tags, electrophoresis and optical detection via laser. The next step in the history of DNA sequencing was to significantly improve throughput while sacrificing accuracy but still trying to retain "good enough" accuracy.

"Sequencing by synthesis" involves taking a single strand of the DNA to be sequenced and then synthesizing its complementary strand.

The "pyrosequencing" method, invented at Stockholm's Royal Institute of Technology by Mostafa Ronaghi, Pål Nyrén, Mathias Uhlén et al.,[313] is based on detecting the activity of DNA polymerase with another chemiluminescent enzyme. Essentially, the method allows sequencing of a single strand of DNA (ssDNA) by synthesizing the complementary strand along it, one base pair at a time, and detecting which base was actually added at each step. Solutions of A, C, G, and T nucleotides are sequentially added to ssDNA and then removed from the reaction. Light is produced only when the nucleotide solution complements the first unpaired base of the template, i.e., enters into a chemical reaction that releases energy leading to a visible signal that can be detected. As shown in Fig. 7.2, both pyrophosphate and a hydrogen ion are released when the fitting nucleotide is added. The proposed pyrosequencing method relies on the detection of pyrophosphate and was first commercialized by the 454 Life Sciences company founded by Jonathan Rothberg.[314]

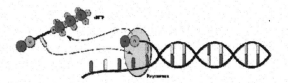

Polymerase integrates a nucleotide.

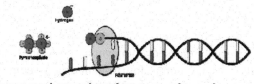

Hydrogen and pyrophosphate are released.

Figure 7.2 Pyrosequencing leads to release of pyrophosphate $(P_2O_7)^{4+}$ and hydrogen ion H^+ (source: Wikipedia).

The winning second-generation sequencing technology, also based on "sequencing by synthesis (SBS)," was developed by Solexa, later (in 2007) to be acquired by Illumina,[315] located in San Diego, CA, the DNA sequencing market leader.

Sample preparation is a very important first step: DNA molecules and primers are first attached on a slide and amplified with polymerase so that local clonal DNA colonies, later coined "DNA clusters," are formed. This method for in vitro clonal amplification is also referred to as "bridge PCR." Then, to determine the sequence, four types of reversible terminator bases (RT bases) are added and non-incorporated nucleotides are washed away. After having taken images of the fluorescently labeled nucleotides, the dye is chemically removed from the DNA, allowing the next cycle to begin. The DNA fragments are extended one nucleotide at a time, allowing image acquisition to be performed at a delayed moment. A key Solexa advantage over 454's pyrosequencing is provided by the decoupling of the enzymatic reaction from the image capture, allowing optimal throughput and theoretically unlimited sequencing capacity.

Whereas second-generation sequencing technologies are relying on both fragment amplification and optical detection, third-generation technologies only use one of those two features.

Pacific Biosystems (PacBio)[316] was founded by Stephen Turner in 2004 with the goal of developing a single-molecule real-time (SMRT) approach for nucleic acid sequencing. The concept, to "eavesdrop on single molecules of DNA polymerase synthesizing virgin DNA in real time,"[317] was initially developed by Turner and his colleague Jonas Korlach at Cornell University in the Laboratories of Watt Webb and Harold Craighead.

The ion torrent concept is based on a combination of DNA fragment amplification and electrical detection, replacing the optical detection of pyrophosphate used in pyrosequencing with electrical detection of the hydrogen ions (protons) which—as shown above in Fig. 7.2—are also released whenever a nucleotide is integrated.

Finally, we may define fourth-generation sequencing as single molecule sequencing with electrical detection.

An example of fourth-generation sequencing technology is provided by Oxford Nanopore.[318]

Since the first decade of the 21st century, the term "Big Data Analytics" has been applied to the analysis of huge volumes of data generated by (among others) retail and financial industries, to understand trends in customer behavior and to design successful marketing campaigns. In the life sciences, the dawn of Big Data already happened in 1953, when Watson and Crick discovered the structure of DNA and the associated role of DNA to store crucial information. Ever since, and accelerated by the Human Genome Project, biologists are developing Big Data Analytics methods for sequencing data.

A typical DNA or RNA sequencing project can be broken down into four principal activities:

1. Sample preparation, experimental design
2. Sequencing
3. Data reduction and management
4. Downstream analysis

The various applications of sequencing and their impact on society and various sectors of the economy are illustrated in Fig. 7.3 (adapted from a Battelle Memorial Institute Report[319]):

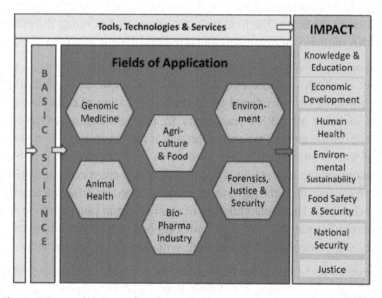

Figure 7.3 Applications of sequencing and impact of the Human Genome Project (adapted by Michael H.).

The completion of the Human Genome Project was a major milestone for the life sciences, but in many ways, it was just the beginning of our human quest to understand life. Instead of answering all biological questions, it created a basis for asking new questions such as how to bridge the gap between a complete genome and the cells and organs of a living organism.

The Human genome's impact on biology is one of a revelation of complexity and will challenge scientists for decades to come. Another important outcome of the HGP was the realization that biology is an information science, giving rise to new disciplines such as bioinformatics, computational biology and systems biology.

Here is a very brief overview of the new fields of investigation that sometimes are referred to as the "omics":

Genomics is the genetic discipline that is focused on sequencing, assembling, and analyzing the function and structure of genomes, defined as the complete set of DNA within a single cell of an organism. Advances in genomics have triggered a revolution in discovery-based research to understand even the most complex biological systems such as the brain. The field includes efforts to determine the entire DNA sequence of organisms and fine-scale genetic mapping.

Transcriptomics is the study of the *transcriptome*, defined as the set of all RNA molecules, including mRNA, rRNA, tRNA, and other noncoding RNA produced in one or a population of cells. Because it includes all mRNA transcripts in the cell, the transcriptome reflects the genes that are expressed at any given time. The study of transcriptomics, also referred to as gene expression profiling, examines the expression level of mRNAs in a given cell population, often using high-throughput techniques based on DNA microarray technology. The transcriptomes of stem cells and cancer cells are of particular interest: transcriptomics applied to those cells help us in the understanding of cellular differentiation and carcinogenesis.

Proteomics is the large-scale study of proteins as encoded by a genome. Proteomics is different from "protein science," which is mostly focused on the 3D structure of proteins and their function. Proteomics generally refers to the large-scale experimental analysis of proteins, it is often specifically used for protein purification and mass spectrometry, an analytical chemistry technique that helps identify the amount and type of chemicals present in a sample by measuring the mass-to-charge ratio and abundance of gas-phase ions. The *proteome* is the entire set of proteins, produced or modified by an organism or system. After genomics and transcriptomics, proteomics is the next step in the study of biological systems. While an organism's genome is more or less constant, the proteome differs from cell to cell and from time to time. Distinct genes are expressed in different cell types, which means that even the basic set of proteins that is produced in a cell requires identification.

Metabolomics is the scientific study of chemical processes involving metabolites, the systematic study of the unique chemical fingerprints that specific cellular processes leave behind. The metabolome represents the collection of all metabolites in a biological cell, tissue, organ, or organism, which are the end products of cellular processes. While mRNA gene expression data and proteomic analyses do not tell the whole story of what might be happening in a cell, metabolic profiling can give an instantaneous snapshot of the physiology of that cell. A related term sometimes used and of great importance in toxicology is *metabonomics*. Metabonomics extends metabolic profiling to include information about perturbations of metabolism caused by environmental factors (including diet and toxins), disease processes, and the involvement of extragenomic influences, such as gut microflora. As to instrumentation, the field of Metabolomics is usually relying on mass spectrometry, whereas metabonomics is more associated with NMR spectroscopy.

Metagenomics (also named environmental genomics, ecogenomics, or community genomics) is the study of genetic material recovered directly from environmental samples. Recent studies use "shotgun" Sanger sequencing or second generation pyrosequencing to get largely unbiased samples of all genes from all the members of the sampled communities. Metagenomics offers a powerful lens for viewing the microbial world. However, a challenge that is present in human microbiome studies is to avoid including the host DNA in the study.

Epigenetics is defined as the study of heritable changes that are not caused by changes in the DNA sequence, but by functionally relevant changes to the genome that do not involve a change in the nucleotide sequence. Epigenetics addresses the question why many cell types of the body maintain drastically different gene expression patterns while sharing exactly the same DNA?

Now let's turn to another genetic engineering technology, that we already touched upon when we discussed the mouse. A big wave of biotech breakthroughs was based on the concept of *monoclonal antibodies*. Paul Ehrlich shared the 1908 Nobel Prize in physiology or medicine[320] for his research into "immunity." Our body's immune system is making "antibodies," proteins that recognize and then try to destroy invading microbes and other attacking organisms.

Monoclonal antibodies (mab) are specific antibodies that are all clones of a unique parent cell, and are made by identical immune cells. The fact that monoclonal antibodies do specifically bind to a given substance, can be used to both detect or to purify that substance.

76 years after Ehrlich, Köhler, Milstein, and Jerne won the Nobel Prize in physiology or medicine for their development of methods to actually produce monoclonal antibodies.[321] They were able to combine the capacity of tumor cells to proliferate forever with a technique to identify antibody-producing cells. These cells were then fused with tumor cells from another species (e.g., mouse), thereby creating hybrid cells with "eternal life" and the capacity to produce the very same antibody in high volumes. They called these hybrid cells *hybridomas*, and as all cells in a given hybridoma come from one single hybrid cell, the antibodies are "monoclonal." The first application of this breakthrough technology was to use a myeloma cancer cell line (with no ability to secrete antibodies) and to fuse it with healthy antibody-producing B cells. Monoclonal antibody (mab) technology could be used to grow huge quantities of pure antibodies aimed at specific selected targets. This led to new diagnostic tests and treatments. By injecting mab's into the bloodstream, the antibodies head straight to their disease target.

Drugs based on monoclonal antibody technology are all named xxxx-mab. The first drug that received FDA approval was *Rituxan* (Rituximab) for the treatment of non-Hodgkin's lymphoma, in 1998. Rituximab is a monoclonal antibody against the protein CD20, which is primarily found on the surface of immune system B cells. Rituximab destroys B cells and is therefore used to treat diseases which are characterized by excessive numbers of B cells, overactive B cells, or dysfunctional B cells. This includes many lymphomas, leukemias, transplant rejection, and autoimmune disorders. In parallel, Genentech developed *Herceptin* (trastuzumab), another monoclonal antibody therapeutic, for the treatment of breast cancer associated with the HER2/neu receptor.

Remicade (Infliximab) was approved by the FDA for the treatment of psoriasis, Crohn's disease, ankylosing spondylitis (an inflammatory disease that can cause vertebrae in the spine to fuse together and causing a hunched-forward posture), psoriatic arthritis, rheumatoid arthritis, and ulcerative colitis.

Humira (Adalimumab) has been approved for rheumatoid arthritis, psoriatic arthritis, ankylosing spondylitis, Crohn's disease, ulcerative colitis, moderate to severe chronic psoriasis and juvenile idiopathic arthritis. Adalimumab was the first fully human monoclonal antibody drug approved by the FDA and is still one of the most prescribed drugs.

Avastin (Bevacizumab) is a recombinant humanized monoclonal antibody that inhibits angiogenesis by targeting vascular endothelial growth factor A (VEGF-A). Angiogenesis, the physiological process through which new blood vessels form from pre-existing vessels, is a normal and vital process in growth and development, as well as in wound healing and in the formation of granulation tissue. However, it is also a fundamental step in the transition of tumors from a benign state to a malignant one.

Genetic engineering can also be applied to the development and production of *vaccines*. The principle is to use genetic engineering technology to isolate the protective antigen gene in the pathogen, and introduce it into the prokaryotic or eukaryotic system to express it and make it into a vaccine. Genetic engineering vaccines are different from vaccines prepared by traditional methods. Because they only contain antigens of pathogenic bacteria or viruses, they are safer and more effective. Since the 1980s, genetic engineering technology has been applied to several targets, with the hepatitis B vaccine as the most successful example.

Another genetic engineering application is *gene therapy*. Gene therapy was originally proposed for single gene deficient genetic diseases. The goal was to have a normal gene to replace the defective gene. Another chosen approach was to somehow eliminate the pathogenic factor of the defective gene. After a lot of trial and error, gene therapy is used today to treat diseases such as liver cancer, hemophilia, hepatitis B, ovarian cancer and other forms of cancer.

So far, we have primarily focused on genetic engineering as applied to human health. Let's now expand our view and explore other opportunities for genetic engineering.

Environmental Protection

Earth's self-cleaning ability has failed to keep pace with human emissions of pollutants. In recent years, the world has paid more and more attention to environmental protection. Due to population growth and economic development, humans have been guilty of polluting the Earth's environment. Several accidents have happened where oil leaked into the ocean. Pristine golden sand beaches were covered by black oil and many seabirds were killed because their feathers were bonded together. Many marine creatures starved to death due to lack of food. The rate at which oil is degraded by nature is very slow because ordinary bacteria can only degrade certain hydrocarbons in petroleum. However, genetically engineered "super bacteria" can decompose most of the hydrocarbons in petroleum. By planting these super bacteria on the beach, it is now possible to effectively degrade oil. In another application, such specialized super bacteria can even devour and transform highly toxic metals such as mercury and cadmium.

Genetic engineering technology can add microorganisms to improve nature's ability to purify the environment. Reduction of soil contamination with heavy metals is one of the most serious global challenges. By means of phytoremediation (use of living plants to clean up soil, air, and water), it is possible to take up, translocate, and transform heavy metals, as well as to limit their toxicity. Genes for phytoremediation may originate from a microorganism, or may be transferred from one plant to another variety better adapted to the environmental conditions at the cleanup site. Organisms that can uptake extremely high amounts of contaminants from the soil are called hyperaccumulators.[322] Phytoextraction can also be performed by plants (e.g., Populus and Salix) that take up lower levels of pollutants, but due to their high growth rate and biomass production, may remove a considerable number of contaminants from the soil.[323]

PCBs are synthetic organic chemical compounds of chlorine and biphenyl. Biphenyl is an aromatic hydrocarbon with molecular formula $(C_6H_5)_2$. It is used as the starting material for the production of polychlorinated biphenyls (PCBs). They have been widely used as dielectric and coolant fluids in many electrical products. Unfortunately, they are toxic. Although no longer in use—they were banned in the 1970s—they are still contaminating rivers, lakes, and harbors and keep posing a threat to human health and to the environment. Previously, dredging and landfilling the PCBs has been the only effective remedy. An example is the ongoing project to dredge the upper Hudson River in New York State. Use of genetic engineering to culture PCB dechlorinators could pave the way for alternative, and possibly more effective, methods of degrading PCBs on-site: *Dehalococcoides mccartyi* is a tiny, strictly anaerobic bacterium that must derive its energy for growth by removing chlorines from chlorinated organic molecules and using them as electron acceptors for respiration, a process known as organohalide respiration.[324] Along with two other bacteria of the Dehaloccoides family and seven known enzymes that act on chlorinated compounds, *D. mccartyi* could be used for bioremediation. Several research teams[325] are working on new and effective ways to culture those microbes.

During the years, a considerable amount of industrial waste has been directly discharged into the environment. Without appropriate treatment, this waste has caused considerable pollution. Now, through genetic engineering technology, the waste can be converted into renewable resources or secondary energy through recycling and re-production. This is an effective measure to prevent environmental pollution and reuse of resources. It is hoped that genetic engineering can be used to identify and to implant desired genes into easily-culturable microorganisms (such as *E. coli*) so that they can be expressed and replicated.

Food and Energy

The Green Revolution, or Third Agricultural Revolution, refers to a set of agricultural technology transfer initiatives during the 1950s and -60s, driven by high-yielding varieties (HYVs) of seeds, chemical fertilizers, and agro-chemicals, irrigation techniques, and new methods of cultivation and mechanization. This initiative is credited with saving over a billion people from starvation, but also had some negative side effects such as the overuse of pesticides.

Genetic engineering may help with the ongoing transformation of green food resources and the improvement of green food quality. Basic resources are plant-based food raw materials, animal-based food raw materials and microbial food raw materials. Raw materials for plant foods promise to increase production, resist pests and diseases. Animal raw materials are mainly cited for the development of superior animal breeds. The application of transgenic technology can improve results of animal breeding. The resource modification of food raw materials mainly uses genetically modified microorganisms to convert cellulose and hemicellulose to produce fungal protein foods. By improving the quality of green foods, we can increase the nutritional value of foods, reduce harmful ingredients in natural foods, and reduce the use of pesticides and fertilizers. For example, to improve the quality of vegetable oils, some genes that are important in the metabolism of plant oils have been cloned, thereby rebalancing the properties of oils and fats.

In the raw material-oriented food industry, compared with traditional mutation breeding, genetic engineering has shown great promise in the large-scale production of food additives such as amino acids, garnishes, and sweeteners. Recombinant microorganisms that highly express secretable amylases, cellulases, lipases, proteinases and other enzyme preparations have also demonstrated their capabilities in food manufacturing, textile printing, leather processing, and daily necessities production. The enantiomers (molecules that are mirror images of each other but not identical— just as left and right hands), which are difficult to separate by traditional chemical methods, can be effectively separated by genetic engineering bacteria.

Energy is an important factor that restricts human production and life. Increasing the utilization rate of traditional energy sources represented by oil, and the industrialization of renewable energy sources, are the hope for solving the energy crisis. In the near future, new types of microorganisms that are constructed using DNA recombination technology, are expected to significantly increase the rate of secondary oil extraction and utilization, to break down hard-to-use cellulose into glucose, and to convert solar energy into chemical and thermal energy.

Future

Genetic editing has recently occupied the media headlines.[326] Many experts and observers predict that gene editing technology, in particular CRISPR, will fundamentally change biology as we know it, change the society we live in and our surrounding plants and animals. Compared to other tools used for genetic engineering, CRISPR (also known as CRISPR-Cas9) is more accurate, cheaper, easier to use, and more powerful. Table 7.1 shows the story of CRISPR, in chronological order and geographic distribution.

Table 7.1 is based on the excellent paper by Eric Lander, published in 2016,[327] and on additional literature reviews by the authors.

There are two parts to the story, first the fascinating discovery of *microbial immunity*, starting with the accidental discovery of "interrupted clustered repeats" in *E. coli* by Y. Ishino et al. of Osaka University in 1987.[328,329] The function of those "palindromic repeats" was not known at the time.

Table 7.1 Story of CRISPR (1987–2013)—from microbial immunity to gene editing

Year	Event/finding	Organism	Scienst(s)	Affiliation	Country
1987	accidental Cloning part of CRISPR	Escherichia Coli	Yoshizumi Ishino	Osaka Univ.	Japan
1993–95	found palindromic repeat sequence of 30 bases + spacer of 36 bases	Haloferax mediterranei + volcanii (Archaea)	Francisco Mojica	Univ of Alicante	Spain
1993	cluster of interrupted direct repeats	Mycobact. tuberculosis	JD van Embden et al.	Health & Environm.	Netherlands
1993	diversity of repeat-intervening sequences	Mycobact. tuberculosis	JD van Embden et al.	Health & Environm.	Netherlands
2000	found CRISPR loci in 20 different microbes	tuberculosis, plague, ...	Francisco Mojica	Univ of Alicante	Spain
2001	Clustered Regularly Interspaced Short Palindromic Repeats (CRISPR)	CRISPR Acronym agreed	F. Mojiva & R. Jansen	Alicante + Utrecht	Spain + Netherlands
2002	Identify genes associated with DNA Repeats	in silico (Bioinformatics)	Ruud Jansen et al.	Utrecht	Netherlands
2003–05	CRISPR is an adaptive immune system	Bioinformatics (BLAST)	Francisco Mojica	Univ of Alicante	Spain
2005	CRISPR: memory of past genetic aggressions	Plague bacteria	C. Pourcel & Vergnaud	Govt + U. Paris Sud	France

Year	Description	System	Researchers	Institution	Country
2005	speculation how CRISPR confers immunity	24 strains Streptococcus t & v	Alexander Bolotin et al.	Govt Agricult. Res.	France
2006–07	Exp. Adaptive immunity, Cas genes 7 & 9	Streptococcus thermophilus	Philippe Horvath et al.	Danisco/Dupont	France/Canada
2006	Cas9 protein active component of immune syst	in silico (Bioinformatics)	Eugen Koonin et al.	NCBI (NIH)	USA
2008	Programming CRISPR - role of Cascade	E. coli	J. van der Oost et al.	Wageningen Univ	Netherlands
2008	CRISPR targets DNA --> Gene editing!	Staphylococcus epidermis	Maraffini, Sontheimer	Northwestern U	USA
2008	Cas9 nuclease cutting DNA at precise positions	Streptococcus thermophilus	Horvath, Barrangou et al.	Danisco/Dupont	USA/France
2010	90% of Archaea and 50% of Bacteria use CRISPR	Review Microbial Immunity	P. Horvath & R. Barrangou	Danisco/Dupont	USA/France
2011	trans-activating CRISPR RNA (tracrRNA) is essential for Cas9 to cleave DNA	NexGen Sequencing and Bioinformatics	Emmanuelle Charpentier and Joerg Vogel, et al.	Vienna, Umeå Berlin, Wuerzburg	Austria, Sweden, Germany
2011–12	reconstitute full CRISPR system in a distant system and *in vitro*	S. thermophilus --> E. coli	Virginijus Siksnys + Horvath & Barrangou	Vilnius Univ + Danisco	Lithuania, France, Wisconsin/USA

(Continued)

Table 7.1 (*Continued*)

Year	Event/finding	Organism	Scienst(s)	Affiliation	Country
2012	Cas9 (+crRNA, tracrRNA) can cut DNA in vitro	E. coli/in vitro	Charpentier & Doudna	Berkeley, Umeå	USA, Sweden
2012	Cas9–crRNA ribonucleoprotein complex mediates specific DNA cleavage	S. thermophilus/ in vitro	Siksnys, Horvath, Barangou, Gasunias	Vilnius Univ + Danisco	Lithuania, France, Wisconsin/USA
2011–12	CRISPR for mammalian (Mouse, human) cells	S. pyogenes, S. thermophilus	F. Zhang	MIT/Broad Inst.	Boston/USA
2013	CRISPR-Cas9 used for human genome editing	human cells	G. Church et al.	Harvard	Boston/USA
2013	Mammalian genome editing	mammalian cells	F. Zhang, Maraffini et al.	MIT/Broad Inst.	Boston/USA
2014	Biology is transformed by facile genome engineering in animals and plants using RNAprogrammable CRISPR-Cas9, a new "Technology for innovative applications in biology"	CRISPR genome editing Review Paper	Jennifer A. Doudna & Emmanuelle Charpentier	UC Berkeley and Howard Hughes/Umeå Univ. Helmholtz Inst.	USA, Sweden and Braunschweig, Germany

In 1993 researchers at the Netherlands' Institute of Public Health and Environmental Protection who investigated *Mycobacterium tuberculosis* noted a cluster of interrupted direct repeats (DR) in this bacterium.[330] They recognized the diversity of the DR-intervening sequences among different strains of *M. tuberculosis* and used this property to design a typing method that was named spoligotyping.[331]

At the same time, repeats were observed in the archaeal organisms of Haloferax and Haloarcula species, and their function was studied by Francisco Mojica at the University of Alicante in Spain. Transcription of the interrupted repeats was also noted for the first time.[332] A subsequent survey of the scientific literature identified interrupted repeats in 20 species of microbes as belonging to the same family.[333] In 2001, Mojica and Ruud Jansen, who was searching for additional interrupted repeats, proposed the acronym CRISPR (clustered regularly interspaced short palindromic repeats) to alleviate the confusion caused by the existence of numerous acronyms that were already used to describe those sequences in the scientific literature.[334] Jansen also found that there are multiple coding sequences in the vicinity of the CRISPR sequence, presumably participating in the physiological function of CRISPR, so it was named CRISPR-associated gene (Cas).

In 2003, Mojica applied bioinformatics to his data and arrived at the fundamentally important conclusion that "CRISPR is an adaptive immune system" for microbes (archaea and bacteria).[335] He struggled for two years to get this breakthrough discovery published: the paper was rejected in 2003 by *Nature*, in 2004 by *PNAS* and by *Molecular Microbiology* and *Nucleic Acid Research*, before it was finally accepted and published (2005) by *Journal of Molecular Evolution*.

Almost simultaneously, through systematic analysis of CRISPR intervening sequences, it was unexpectedly found that they originated from viruses or plasmids, defined as genetic structures in a cell that can replicate independently of the chromosomes (typically a small circular DNA strand in the cytoplasm of a bacterium). Eugene Koonin, an evolutionary biologist at the National Center for Biotechnology Information (NCBI), realized that bacteria can use CRISPR as an important weapon against virulence.[336] This sparked interest in the French microbiologist Rodolphe Barrangou, who worked for the yogurt company Danisco (later acquired by

Dupont group). Led by Philippe Horvath, Danisco first confirmed experimentally in 2007 that CRISPR-Cas is a bacterial-acquired immune system.[337] In 2008, Horvath et al. discovered the important role played by "CRISPR associated (Cas) proteins," they found that "Cas9 nuclease is cutting DNA at precise positions."[338] In 2010, Horvath et al. published (in *Science*) an excellent review[339] in which it was stated that *90% of archaea and 50% of bacteria use CRISPR for their immune defense*. Barrangou's and Horvath's contributions are now universally recognized.[340] It should also be said that the French government[341,342] was involved in the research that eventually led to the insights described above. That's no surprise given the strong French bacterial research tradition created by Pasteur and his disciples.

The second part of the CRISPR story is related to the translation of the microbial immunity discovery into a gene editing technology. It's one of the most compelling examples of how sapiens can derive great benefits from the study of advanced evolutionary developments— like the way microbes defend themselves against a virus or plasmid!

CRISPR-related *gene editing* can be traced back to attempts by Marrafini and Sontheimer,[343] who confirmed that "CRISPR targets DNA" and suggested that the CRISPR mechanism could be used for gene editing. They even tried to patent this idea but failed to convince the US patent office due to insufficient experimental evidence. A missing piece was then found by Emmanuelle Charpentier and Joerg Vogel, who in 2011 published an article in *Nature*, which first demonstrated that trans-activating CRISPR RNA (tracrRNA) is essential for Cas9 to cleave DNA.[344]

Another important step towards "CRISPR programming" was taken by Virginijus Siksnys (Vilnius, Lithuania) and by Danisco's Horvath and Barrangou, who collaborated on a first in vitro CRISPR system.[345] Almost simultaneously, Charpentier and Jennifer Doudna, a structural biologist at the University of California, Berkeley, worked together on the demonstration that "Cas9 +crRNA + tracrRNA can cut DNA in vitro." The results of the collaboration between Charpentier and Doudna were published in June 2012.[346] Independently, both research teams had achieved an important scientific breakthrough: the CRISPR-Cas9 system can be used to achieve target DNA cleavage for genetic editing purposes. Charpentier and Doudna further re-engineered the Cas9 endonuclease into a more manageable two-component system by fusing the two RNA molecules into a "single-guide RNA" (sgRNA) that, when combined with Cas9, could find and cut the DNA target specified by the guide RNA. By manipulating the nucleotide sequence of the guide RNA, the artificial Cas9 system could be programmed to target any DNA sequence for cleavage.

Figure 7.4 illustrates the described "translation" of microbial immunity to a new revolutionary gene editing technology.

In Fig. 7.4, protospacer adjacent motif (PAM) is a 2–6 base pair DNA sequence immediately following the DNA sequence targeted by the Cas9 nuclease in the CRISPR bacterial adaptive immune system. PAM is a component of the invading virus or plasmid, but is not a component of the bacterial CRISPR locus. Cas9 will not successfully bind to or cleave the target DNA sequence if it is not followed by the PAM sequence.

The rest of the CRISPR story includes heavy involvement by MIT and Harvard University, along with the Broad Institute.[348] At the beginning of 2013, Feng Zhang[349,350] and George Church[351] of Harvard University edited specific genes in mammalian cells, especially using multiple sgRNAs to achieve simultaneous multi-gene knockout, which greatly improved editing efficiency and the scope of application. In addition, Zhang's lab confirmed that Cas9 can accurately cut endogenous loci in human and mouse cells under the guidance of small RNA. The research team of Lei S. Qi (now at Stanford University) at the University of California, San Francisco, System and Synthetic Biology Center invented the CRISPRi technology to achieve effective inhibition of genes in *E. coli* without significant off-target effects. Moreover, it is possible to simultaneously inhibit multiple genes in mammalian cells. Genetic editing of zebrafish, fungi, and bacteria was quickly implemented using the CRISPR/Cas system. In May 2013, the Jaenisch team (Whitehead Institute) used CRISPR/Cas-mediated genetic engineering to create mice with multiple mutations in multiple genes,[352] greatly facilitating functional studies of multiple genes in vivo. Subsequently, and as illustrated in Fig. 7.5, genetic editing of various organisms such as fruit flies, nematodes, rats, pigs, sheep, rice, wheat, and sorghum was carried out, setting off a research boom that continues to this day.

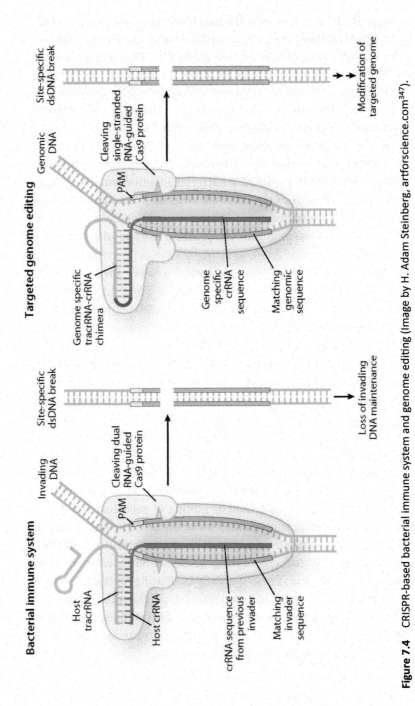

Figure 7.4 CRISPR-based bacterial immune system and genome editing (Image by H. Adam Steinberg, artforscience.com[347]).

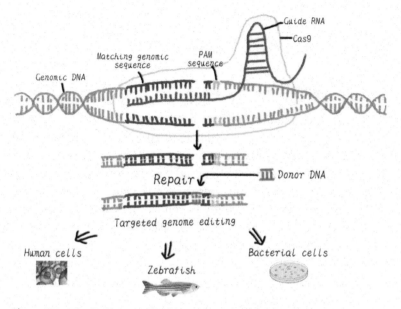

Repair

Figure 7.5 The basic principle of gene editing technology (source: BGI).

What are the pros and cons of CRISPR for gene editing?

The advantages of CRISPR are obvious: CRISPR removes the previous dependence on embryonic stem (ES) cells and enables the editing of a variety of species. CRISPR technology relies on Cas9 and sgRNA to achieve "precision" cutting, simplifying the experimental steps and greatly shortening the experimental cycle. The CRISPR cycle is short and the workload is small, which is followed by a big reduction in costs.

However, CRISPR also has some technical challenges: sgRNA and Cas9 protein are not always working accurately. Even when sgRNA is precisely positioned, it is not always reliable. The Cas9 protein should recognize the standard PAM sequence very well, but it may occasionally be off-target, which affects the editing efficiency.

CRISPR's biggest challenge, however, may be related to bioethics and to attempts to regulate CRISPR-based research. As pointed out by Heidi Ledford[353] in her *Nature* paper, regulation will have a hard time catching up to the widespread use of this technology. As she is pointing out, "it will be hard to detect whether something has been mutated conventionally or genetically engineered" via CRISPR.

Below we list some of the most exciting applications of this revolutionary technology and the obstacles that may slow or prevent it from fulfilling its full potential.

CRISPR Applications

There is a huge unmet medical need caused by the shortage of human organs for transplantation. In Chapter 6B, we discussed the pig's physiological similarity to humans. In 2015, Luhan Yang and George Church co-founded eGenesis[354] to create "PERV-free pigs." PERV (porcine endogenous retrovirus) is causing pig-to-human and human-to-human infectivity in cell cultures, but it was shown recently that CRISPR can be used to successfully remove PERV—a key requirement for xenotransplantation.

CRISPR can correct genetic errors that cause disease

Hypertrophic cardiomyopathy (HCM) is a heart disease that affects approximately 1 in 500 people worldwide. This disease is painful and often fatal. Some dominant gene mutations can cause hardening of the heart tissue, which may lead to chest pain, weakness, and even severe cardiac arrest. Thanks to recent medical advances, the average life expectancy of HCM patients is close to that of the average person, but if not treated in time, it can lead to life-threatening conditions.

But in the future, we may be able to cure this disease completely through genetic editing. In the summer of 2017, scientists at Oregon Health and Science University used CRISPR to remove a defective gene from human embryos.[355]

Another recent breakthrough has been reported by Chinese scientists who used a variation of CRISPR to correct a single amino acid that causes Marfan syndrome, a connective tissue disorder that causes problems with the cardiovascular system, the skeletal system and the eyes of patients. The researchers[356] did not detect any off-target effects of CRISPR or unwanted insertions or deletions in the genomes of the seven modified embryos, none of which was implanted.[357]

The results are exciting: most of the positive embryos using CRISPR-Cas9 technology within 18 hours of fertilization did not have a gene mutation (almost no chance of developing this disease). In other words, by using this technology, we may reduce the burden of this genetic disease on the family and the entire human population. Catching mutations in the early stages of embryonic development can reduce or even eliminate the need for treatment later in life.

CRISPR can be used to create same-sex offspring

When describing the Komodo dragon, we already discussed parthenogenesis, a natural form of asexual reproduction in which growth and development of embryos occur without fertilization. Parthenogenesis means development of an embryo from an unfertilized egg cell.

Scientists in China have been able to create offspring in mice from both two mothers and from two fathers.[358] For the first time, researchers have used the DNA from two mouse mothers to create healthy pups, some of which matured and had their own offspring. The scientists also produced baby mice using the combined genetic material from two fathers, although those pups only lived for a couple of days.

Aside from ethical issues, scientists are still skeptical that the technique used in mice could ever be applied to people. How was it done in mice?

In all mammals, scientists have found "genetic imprints," small chemical tags that attach to DNA and turn off a gene. There are about 100 known such tags, primarily on genes affecting an embryo's growth. Many genes that are tagged in one sex remain untagged in the opposite sex. However, it is not possible to combine two genes that are tagged in an embryo—which would happen with parents of the same sex. If both genes are tagged the embryo would die.

The answer to this problem is to "delete tags": study author Qi Zhou, a developmental biologist at the Chinese Academy of Sciences in Beijing, and his team used lab-grown embryonic stem cells from either a sperm or an egg. These cells have only one set of chromosomes and, like most cells, contain genetic regions that can produce the chemical tags. Using CRISPR, Qi Zhou and his team deleted these genetic regions in batches, searching for groups that could be removed without stopping the production of a healthy embryo. The team then combined a stem cell from a female mouse with the egg from another female to create pups from two mothers. They also took a stem cell from a male and injected it, along with another male's sperm, into an egg without a nucleus to create offspring from two fathers.

After deleting three genetic regions, the scientists managed to produce 29 living mice from two females, 7 of which went on to have their own pups. The team needed to delete 7 regions to produce 12 pups from two male parents—but those baby mice lasted only 2 days after succumbing to problems including trouble breathing and extra fluid in their tissues.[359] According to Zhou, the study showed "a new and clear way to produce offspring between same-sex mammals."

CRISPR can eliminate microorganisms that cause disease

Although the treatment of AIDS has turned the virus from a deadly killer into a chronic health condition, scientists have yet to find a cure. This situation may change as CRISPR technology advances. In 2017, a Chinese research team successfully enhanced the resistance of mice to HIV by replicating a gene mutation that effectively prevented the virus from entering the cell.[360] At present, scientists only perform these experiments on animals, but there is reason to believe that the same method also applies to humans. HIV resistance mutations occur in a small number of people. By using CRISPR to introduce mutations into human stem cells, researchers could significantly increase human resistance to AIDS in the future.

RECENT NEWS: on November 25, 2018, the young Chinese researcher Jiankui He became the center of a global firestorm when it emerged that he had overstepped an ethical threshold by creating the first CRISPR-edited babies, twin girls named Lulu and Nana.[361]

CRISPR can resurrect extinct Species

In February 2017, the Harvard University geneticist George Church issued a surprising statement at the annual meeting of the American Association for the Advancement of Science (AAAS). He claimed that his team could breed the elephant-mammoth hybrid embryo within 2 years.[362] Church hopes to resurrect woolly mammoths to control global warming. His idea is that the mammoth can help remove the snow from the thawing zone above permafrost, and thereby permit cold air to reduce the risk of thawing (that would release catastrophic amounts of methane.)

Church and his team hope to use CRISPR to combine the elephants of Asia (possibly rescued endangered species) with the genetic material of woolly mammoths. The latter sample was extracted from frozen mammoth DNA found in Siberia. By adding Mammalia's genome to Asian elephants, the final organism will have the common characteristics of long-haired elephants. The long hair can then be used as insulation in cold climates.

CRISPR can create healthier new food

CRISPR gene editing technology has proved very promising in agricultural research. Scientists from the Cold Spring Harbor Laboratory in New York used the tool to increase tomato production.[363] The laboratory developed a method to edit the genes that determine the size of the tomato, the structure of the branches, and ultimately the shape of the plants to gain more. Each feature can be controlled like a light switch. Using native DNA to enhance the natural offerings can help break yield barriers.

The Future of CRISPR

Current scientific progress shows that CRISPR is not only an extremely versatile technology, it has also proven to be accurate and safer to use. But it still has a lot of room for improvement, and we are only just starting to see the full potential of a genome editing tool like CRISPR-Cas9. There are still technical and ethical obstacles in our use of genetically modified foods to meet human needs, eradicate genetic diseases, or revive extinct species, but we still need to follow this direction.

Nanotechnology

When discussing capabilities of specialized animals that supersede human abilities, an obvious next question is to wonder how those capabilities could somehow be added to the list of human assets. Is it just science fiction, or is there a chance to realize some of those dreams?

Humans have always envied birds that can fly without much effort, navigate across large distances, and have much superior vision. What sapiens has done to deal with those impossible dreams was to develop technology and to build devices that can enable what we are not born with.

Recent advances in nanotechnology could push those attempts much further and open up new areas like bionic eyes, stronger bodies and much better resistance to diseases, bone fractures, and muscle problems. As microprocessors are turning into nano-processors, it makes sense to include computers in the definition of nanodevices. Nanotechnology will therefore not only enable future medical instruments and devices of all kinds, but also open the door to more advanced software solutions that will support and improve our clinical decision making and eventually enhance the biggest human asset—our brain.

Summary

By studying animals, we can learn about their respective strengths, and with new technologies such as genetic engineering, gene editing, and nanotechnology, we may have a chance to improve tomorrow what we are today. We have to be careful and observe strict bioethical guidelines, but we will certainly benefit.

Chapter 8

Animal Connection Challenges

Throughout this book, we have focused on what sapiens can learn by studying other species. We took an optimistic and positive point of view and did not focus much on possible health challenges caused by getting too close to animals or otherwise going too far with our interactions.

There are accidents such as snakebites, encounters with sharks at beaches or with big cats or other wild animals such as wolves or bears that can cause problems for humans. What is certainly even more serious is the exposure to insects such as mosquitoes, in particular in tropical regions. Over 1 million people worldwide die from mosquito-borne diseases every year, making those diseases a serious threat to human health. Mosquito-initiated diseases include malaria, dengue, encephalitis and yellow fever. Another potentially serious condition is Lyme disease, which is caused by Borrelia, a genus of bacteria of the phylum Spirochete. Lyme borreliosis is a zoonotic,[a] vector-borne disease transmitted primarily by ticks and by lice[368] that first infest rodents and then deer before jumping on humans and dogs. The genus is named after the French biologist Amédée Borrel (1867–1936).

---

[a] A zoonotic disease normally exists in animals but can infect humans.

Our Animal Connection: What Sapiens Can Learn from Other Species (Second Edition)
Michael Hehenberger and Zhi Xia
Copyright © 2021 Jenny Stanford Publishing Pte. Ltd.
ISBN 978-981-4877-50-3 (Hardcover), 978-1-003-13072-7 (eBook)
www.jennystanford.com

However, what disrupts human civilization more than anything except war is the threat of infectious diseases causing epidemic or even pandemic outbreaks. Since "our animal connections" are often responsible for such infectious disease outbreaks, we will discuss them in some detail below.

A pandemic (or global epidemic) is a disease that affects people over an extensive geographical area. Below we present an overview of pandemics with strong animal connections. We are excluding smallpox, measles and typhus, although it is now a widely held opinion that close to 75% of recently emerging infectious diseases affecting humans are diseases of animal origin, and that approximately 60% of all human pathogens are zoonotic.[369] As a result, many of the diseases that we now think of as distinctly human diseases, such as measles, smallpox, and diphtheria are believed to have originally jumped from animals to man.

Plague

Plague or the "black death" is the disease caused by the bacterium Yersinia pestis. It was most likely carried by *fleas living on black rats* that may have first spread by traveling on merchant ships. However, the major Black Death pandemic described below was in large part spread by human fleas which cause a pneumonic version of the plague. Pneumonic plague enables person-to-person infection via aerosols, thus explaining the very quick inland spread of the epidemic, which was faster than would be expected if only caused by rat fleas.

- From 541 to 750, the first plague pandemic, starting with the Plague of Justinian, killed between 50% and 60% of Europe's population.
- The Black Death (1347–1352) killed 25 million people in Europe. This second plague pandemic reduced the world population by 75–100 million, from about 450 million to 350–375 million.

Influenza

- The first European influenza epidemic occurred between 1556 and 1560, with an estimated mortality rate of 20%. The

Influenza Pandemic of 1918[370] killed 25–50 million people (about 2% of world population of 1.7 billion). It affected Germany, the United Kingdom, France, and the United States during World War I. The name *Spanish flu* is a serious misnomer: To keep morale high, newspapers in the countries that participated in WWI, were not allowed to report the epidemic. Only in Spain, a country that had remained neutral, newspapers were free to report the epidemic's effects, such as the grave illness of King Alfonso XIII. These stories created a false impression of Spain as having been especially hard hit. Since then, it has been speculated (and later decisively confirmed) that this pandemic was triggered by a strain of the influenza virus that originated with humans living closely together in camps and living in the vicinity of farm animals such as hens, ducks, geese and pigs. One theory speculates that the virus strain originated in Kansas, USA, in poultry and swine that were bred for local consumption. The second hypothesis, formed through reconstruction of the virus, suggests that it jumped directly from birds to humans, without traveling through swine. A possible first outbreak happened in a British troop staging camp in Etaples, France. The 1918 Spanish flu was the first of two pandemics caused by H1N1 influenza A virus, to be discussed in more detail below. The second H1N1-related pandemic was the swine flu pandemic of 2009.[371] Even today, influenza kills about 250,000 to 500,000 worldwide each year.

Measures to Reduce Pandemic Risk throughout History

In the mid-19th century, important steps were taken to significantly improve public health and reduce pandemic risk, starting with water quality. Other milestones in the fight against infectious diseases are listed below:

- Louis Pasteur (1822–1895) proved beyond doubt that certain diseases are caused by infectious agents, and developed a vaccine for rabies (which can be transmitted by dogs, foxes, etc.).

- Robert Koch (1843–1910) provided the study of infectious diseases with a scientific basis known as Koch's postulates (see Fig. 8.1) that since then were applied to vaccine development.
- Jonas Salk (1914–1995) and coworkers developed effective vaccines for smallpox and polio.

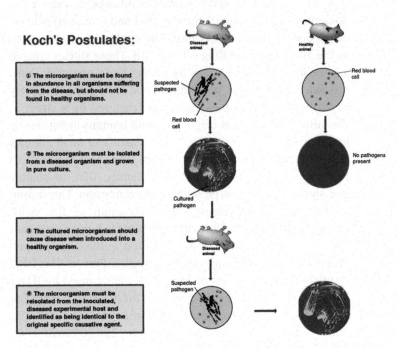

Koch's Postulates:

① The microorganism must be found in abundance in all organisms suffering from the disease, but should not be found in healthy organisms.

② The microorganism must be isolated from a diseased organism and grown in pure culture.

③ The cultured microorganism should cause disease when introduced into a healthy organism.

④ The microorganism must be reisolated from the inoculated, diseased experimental host and identified as being identical to the original specific causative agent.

Figure 8.1   Koch's four postulates (source: Wikipedia).

Koch later abandoned the strong requirement of the first postulate when he discovered *asymptomatic carriers* of cholera and typhoid fever. Asymptomatic or subclinical infection carriers are now known to be a common feature of many infectious diseases, especially viral diseases.

Another important contributor to the improvement of public health was William Osler (1849–1919). A well-known quote attributed to him is "Soap and water and common sense are the best disinfectants."

Due to antibiotics, bacterial diseases are now considered less serious than viral diseases. Human immunity against viral diseases is still a great weakness of our species.

Viral Pathogens and Therapies

A virus is a biological agent that reproduces inside the cells of living hosts. According to our definition of *life*, it is not a form of life because it cannot exist without its host. It uses the host cell's "molecular machinery" to replicate. When infected by a virus, a host cell is forced to produce identical copies of the original virus, usually at an extraordinary rate, resulting in large numbers. Unlike most living organisms (including bacteria), viruses do not have cells that divide; new viruses are assembled in the infected host cell. However, viruses contain genes, which give them the ability to mutate and evolve.

An important concept for viral classification is the "viral envelope." Many viruses have envelopes covering their protective protein *capsid*, i.e., the protein shell of a virus, enclosing the genetic material of the virus. The envelopes typically are derived from portions of the host cell membranes (phospholipids and proteins), but may include some viral glycoproteins.[b] Viral envelopes are essential for providing the entry into host cells. They may even help viruses to avoid the host immune system. Glycoproteins on the surface of the envelope serve to identify and bind to receptor sites on the host's membrane. The viral envelope then fuses with the host's membrane, allowing the capsid and viral genome to enter and infect the host. Figure 8.2 shows the viral envelope for the Cytomegalovirus, a member of the viral family known as Herpesviridae or herpesviruses. Herpesviruses like CMV (frequently associated with the salivary glands) share a characteristic ability to remain latent within the body over long periods.

The cells that are invaded by the virus will often die or be weakened. They shed more viral particles for an extended period. The lipid bilayer envelope of these viruses is quite sensitive to the removal of moisture (desiccation), to heat, and to detergents. Therefore, these viruses are easier to sterilize than non-enveloped viruses, have limited survival outside host environments, and typically must transfer directly from host to host. Enveloped viruses possess great adaptability and are known to change in a short time in order to evade the immune system and cause persistent infections.

---

[b]Glycoproteins are proteins which contain carbohydrates (glycans) covalently attached to amino acid side-chains. Glycoproteins are often important integral parts of cell membranes.

Scheme of a CMV virus

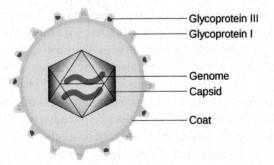

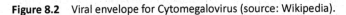

Figure 8.2 Viral envelope for Cytomegalovirus (source: Wikipedia).

Viral infections can cause disease in humans, animals and even plants. However, they are usually eliminated by the immune system, often (but not always) conferring lifetime immunity to the host for that virus. Antibiotics have no effect on viruses, but some antiviral drugs have been developed to treat certain life-threatening infections. *Vaccines* that produce immunity for a limited period of time (months, years, even a lifetime) can prevent some viral infections. A vaccine is providing active *acquired immunity* to a particular infectious disease. It typically contains an agent that resembles a disease-causing microorganism and is often made from weakened or killed forms of the microbe, its toxins, or one of its surface proteins. The agent stimulates the body's immune system to recognize the agent as a threat, to destroy it, and to further recognize and destroy any of the microorganisms associated with that agent that it may encounter in the future. Vaccines can be either prophylactic (to prevent or ameliorate the effects of a future infection by a natural or "wild" pathogen), or therapeutic (to fight a disease, such as cancer).

The origin of the name "vaccine" is related to Edward Jenner's discovery, in 1798,[372] that smallpox of the cow (*Variolae vaccinae*) has a protective effect against smallpox. In 1881, to honor Jenner, Louis Pasteur proposed that the terms "vaccine" and "vaccination" should be extended to cover the new protective inoculations that then were developed. However, the basic idea behind vaccination, to train the immune system with a small dose of a disease pathogen, goes back to Chinese medicine in the 16th century.[373]

Viruses spread in many ways. Plant viruses are often spread from plant to plant by insects and other organisms, known as vectors (a person, animal or microorganism that carries and transmits an infectious pathogen into another living organism). Some viruses of animals, including humans, are spread by exposure to infected bodily fluids. Viruses such as influenza are spread through the air by droplets of moisture when people cough or sneeze. Viruses such as norovirus are transmitted by the fecal–oral route, which involves the contamination of hands, food and water. Rotavirus is often spread by direct contact with infected children. HIV, the human immunodeficiency virus, is transmitted by bodily fluids transferred during sexual relations. Others, such as the Dengue virus, are spread by blood-sucking insects.

Diagnosis

For viral diagnosis, a regular light microscope may not be sufficient. Therefore, microscopy is often used in conjunction with biochemical staining techniques, applied in combination with antibody-based[c] techniques. For example, the use of antibodies that were made artificially fluorescent can be directed to bind to and identify a specific antigen present on a pathogen. A fluorescence microscope[374] is then used to detect fluorescently labeled antibodies bound to internalized antigens within clinical samples or cultured cells. This technique is especially useful in the diagnosis of viral diseases, where the light microscope is incapable of identifying a virus directly.

For viruses that have been genetically sequenced, a genetic test can also be an accurate way of identification.

Virus classification and the importance of animal transmissions

The *Baltimore classification*, developed by Nobel Laureate David Baltimore,[375] is a virus classification system that groups viruses

---

[c]An antibody (Ab) is a large, Y-shaped protein produced mainly by plasma cells that is used by the immune system to neutralize pathogens such as pathogenic bacteria and viruses. The antibody recognizes a unique molecule of the pathogen, called an antigen.

into families, depending on their type of genome (DNA, RNA, single-stranded (ss), double-stranded (ds), etc.) and their method of replication.

As a general rule, *DNA viruses* replicate within the nucleus of the infected host cell, while RNA viruses replicate within the cytoplasm.

DNA virus–related diseases are very common and include herpes, chickenpox, shingles, mononucleosis, smallpox, hepatitis B, etc. More than 90% of adults have been infected with at least one of these, and a latent form of the virus remains in most people.

RNA virus–related diseases include the common cold, influenza, coronavirus related diseases, hepatitis C, hepatitis E, West Nile fever, Ebola virus disease, rabies, polio and measles. Viruses with RNA as their genetic material which also include DNA intermediates in their replication cycle are called retroviruses. Notable human retroviruses include HIV-1 and HIV-2, the cause of the disease AIDS.

Viral diseases also affect animals. In rare cases, they can "jump" from animals to humans and cause challenges hard (or impossible) to handle by our immune system.

The term *zoonotic* pertains to a zoonosis, namely a disease that can be transmitted from animals to people.

Rhabdoviruses are a diverse family of RNA viruses that infect a wide range of hosts, from plants and insects, to fish and mammals. The Rhaboviridae family consists of six genera of which two only infect plants. Novirhabdoviruses infect fish, and vesiculovirus, lyssavirus and ephemerovirus infect mammals, fish and invertebrates. The family includes pathogens such as *rabies* virus, vesicular stomatitis virus and potato yellow dwarf virus that are of public health, veterinary, and agricultural significance. It is well known that humans bitten by mammals infected by the rabies virus develop severe symptoms.

Foot-and-mouth disease virus (FMDV) is a member of the *Aphthovirus* genus in the Picornaviridae family and is the cause of foot-and-mouth disease in pigs, cattle, sheep and goats. It is a non-enveloped, positive strand, RNA virus. FMDV is a highly contagious virus. It enters the body through inhalation.

Pestiviruses have RNA genomes. They cause classical swine fever (CSF) and bovine viral diarrhea (BVD). Mucosal disease is a distinct, chronic persistent infection, whereas BVD is an acute infection.

Arteriviruses are small, enveloped, animal viruses with an icosahedral core containing a positive-sense RNA genome. The family includes equine arteritis virus (EAV), porcine reproductive and respiratory syndrome virus (PRRSV), lactate dehydrogenase elevating virus (LDV) of mice and simian hemorrhagic fever virus (SHFV).

Coronaviruses (see Fig. 8.3) are enveloped viruses with a positive-sense RNA genome and with a nucleocapsid of helical symmetry. They usually infect the upper respiratory and gastrointestinal tract of mammals and birds. They are the cause of a wide range of diseases in cats, dog, pigs, rodents, cattle, other mammals, and humans.

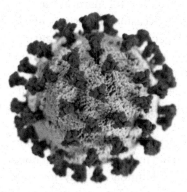

Figure 8.3 Illustration of the morphology of coronaviruses. The genetic part is RNA.

We will cover the Coronavirus and its "animal connection" in much more detail below.

Torovirus is a genus of viruses within the family Coronaviridae that primarily infect vertebrates and include Berne virus of horses and Breda virus of cattle. They cause gastroenteritis in mammals, rarely including humans.

Influenza is caused by RNA viruses of the family Orthomyxoviridae and affects birds and mammals.

Orthomyxoviridae include seven genera: Alphainfluenzavirus, Betainfluenzavirus, Deltainfluenzavirus, Gammainfluenzavirus, Isavirus, Thogotovirus, and Quaranjavirus. As to "our animal connection," the first four genera are of greatest interest. They contain viruses that cause influenza in vertebrates, including birds (see also avian influenza), humans, and other mammals. In addition, isaviruses infect salmon; the thogotoviruses are arboviruses, infecting vertebrates and invertebrates, such as ticks and mosquitoes. The quaranjaviruses are also arboviruses, infecting arthropods as well as birds.

- Alphainfluenzavirus infects humans, other mammals, and birds, and causes all flu pandemics.
- Betainfluenzavirus infects humans and seals.
- Deltainfluenzavirus infects pigs and cattle.
- Gammainfluenzavirus infects humans, pigs, and dogs.

Wild aquatic birds are the natural hosts for a large variety of influenza A viruses. Occasionally viruses are transmitted from this reservoir to other species and may then cause devastating outbreaks in domestic poultry or give rise to human influenza pandemics, such as the Spanish flu already covered above.

Typically, influenza is transmitted from infected mammals through the air by coughs or sneezes, creating aerosols containing the virus, and from infected birds through their droppings. Influenza can also be transmitted by saliva, nasal secretions, feces, and blood. Flu viruses can remain infectious for about one week at human body temperature, but indefinitely at very low temperatures (such as lakes in northeast Siberia). They can be inactivated by disinfectants and detergents.

Influenza viruses have 2 surface glycoproteins, *hemagglutinin* and *neuraminidase*, as shown in Fig. 8.4.

The viruses bind to a cell through interactions between its hemagglutinin glycoprotein and sialic acid sugars on the surfaces of epithelial cells in the lung and throat. Influenza hemagglutinin (HA) is found on the surface of the influenza viruses. It is responsible for binding the virus to cells with sialic acid on the membranes, such as cells in the upper respiratory tract or erythrocytes. The name *hemagglutinin* comes from the protein's ability to cause red blood cells (erythrocytes) to clump together (*agglutinate*) in vitro.

Influenza neuraminidase (also named sialidase) exists as a mushroom-shape projection on the surface of the influenza virus. Sialidase cleaves sialic acid to form new particles. This cleavage releases new viruses which can later invade new host cells. Neuraminidase actions include assistance in the mobility of virus particles through the respiratory tract mucus. Neuraminidase is therefore a target for antiviral drugs, such as Oseltamivir, sold under the brand name Tamiflu.[376]

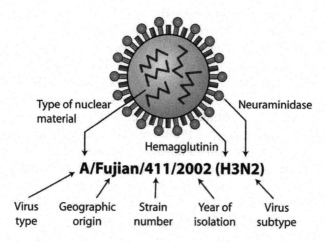

Type of nuclear material

Neuraminidase

Hemagglutinin

A/Fujian/411/2002 (H3N2)

Virus type | Geographic origin | Strain number | Year of isolation | Virus subtype

Figure 8.4 Classification of influenza virus strains by surface proteins "H" and "N."

The type A influenza viruses are the most virulent human pathogens among influenza types and cause the most severe disease. Influenza A viruses are further classified based on their (up to 16) viral surface proteins hemagglutinin (HA or H) and (up to nine) neuraminidase (NA or N), as shown above in Fig. 8.4.

The serotypes, ordered by the number of known human pandemic deaths, are as follows:

- **H1N1** caused "Spanish flu" in 1918 (25–50 million deaths), and the "swine flu" in 2009: analysis of the virus pointed to a combination of bird, swine, and human flu viruses that further combined with a Eurasian pig flu virus,[377] eventually resulting in the term "swine flu." The number of WHO lab-confirmed deaths was 18,449, although more than 10% of the global population was infected. The H1N1/09 virus deviated from the norm by *not* disproportionately infecting adults older than 60 years.

- **H2N2** (A/H2N2) is a subtype of influenza A virus. H2N2 has mutated into various strains including the "Asian flu" strain which now seems to be extinct. However, the 1957–1958 influenza pandemic was a global pandemic which originated in Guizhou, China, and killed at least 2 million people.

- **H3N2** caused "Hong Kong flu": The first recorded instance of the outbreak appeared in July 1968 in Hong Kong. By the end of July 1968, extensive outbreaks were reported in Vietnam and Singapore. By September 1968, the flu had reached India, the Philippines, northern Australia, and Europe. The virus entered California, carried by returning troops from the Vietnam War, but did not become widespread in the United States until December 1968. It reached Japan, Africa, and South America by 1969. In Berlin, the excessive number of deaths led to corpses being stored in subway tunnels. In total, East and West Germany registered 60,000 estimated deaths. In some areas of France, 50% of the workforce was bedridden, and manufacturing suffered large disruptions due to absenteeism. The H3N2 virus returned during the following 1969/1970 flu season, resulting in a second, even deadlier wave of deaths. It remains in circulation today as a strain of the seasonal flu. The United States Centers for Disease Control and Prevention (CDC) estimated that, in total, the virus killed 1 million people worldwide, from its beginning in July 1968 until the winter of 1969–70. About 100,000 people died in the U.S. and the virus affected particularly those 65 years and older.

- **A/H5N1** is a subtype of the influenza A virus which can cause illness in humans and many other animal species, in particular birds. A bird-adapted strain, called **HPAI A(H5N1)** for "highly pathogenic avian influenza virus of type A of subtype H5N1," is causing the H5N1 flu, commonly known as avian influenza ("bird flu"). It has been killing tens of millions of birds and forced the culling of hundreds of millions of others to stem its spread. References to "bird flu" and H5N1 in the popular media refer to this strain. It is easily spread but rarely fatal. In July 2013, the WHO announced a total of 630 confirmed human cases which resulted in the deaths of 375 people since 2003. According to WHO, H5N1 pathogenicity is continuing to rise, but the avian influenza disease situation in farmed birds is being held in check by vaccination.

- **H7N7** has unusual zoonotic potential. H7N7 can infect humans, birds, pigs, seals, and horses in the wild; and has infected mice in laboratory studies. This zoonotic potential represents a pandemic threat. As to human infections, 89 people in the Netherlands were confirmed to have been infected by H7N7 in 2003, following an outbreak in poultry on approximately 255 farms. One death was recorded—a veterinarian who had been testing chickens for the virus—and all infected flocks were culled. Most affected people had only mild symptoms including conjunctivitis. Antibodies were found in over half of 500 people tested according to the final official report by the Dutch government:

- **H1N2** is endemic in humans and pigs. Between December 1988 and March 1989, 19 influenza H1N2 virus isolates were identified in 6 cities in China, but the virus did not spread further. During the 2001–02 Northern hemisphere flu season, H1N2 was encountered in Canada, USA, Ireland, Latvia, France, Romania, Oman, India, Malaysia, and Singapore with earliest documented outbreak of the virus occurring in India. H1N2 is not causing great concern because the seasonal flu vaccine should provide good protection.

- **H9N2** is the most common subtype of influenza viruses in Chinese chickens. Even under the long-term vaccination programs, it causes great economic loss for the poultry industry. As to human infections with avian influenza, H9N2 is playing a role as the gene donor for H7N9 and H10N8 viruses that are infecting humans too. H9N2 viruses have a wide host range. In China, often regarded as a breeding ground of avian influenza viruses, the H9N2 virus has been detected in multiple avian species, including chicken, duck, quail, pheasant, partridge, pigeon, silky chicken, chukar and egret.

Flu vaccines are based on predicting which "mutants" of H1N1, H3N2, H1N2, and influenza B, will proliferate in the next season. Separate vaccines are developed for the Northern and Southern Hemispheres in preparation for their annual epidemics. In the tropics, influenza shows no clear seasonality. In the past ten years, H3N2 has tended to dominate in prevalence over H1N1, H1N2, and influenza B.

Flaviviruses

constitute another family of RNA viruses, including the West Nile virus, dengue virus, tick-borne encephalitis virus, yellow fever virus, and several other viruses. They are typically arthropod borne and can replicate in both mammalian and insect cells. The viruses in this family that are of veterinary importance include Japanese encephalitis virus, St. Louis encephalitis virus, West Nile virus, Israel turkey meningo-encephalomyelitis virus, Sitiawan virus, Wesselsbron virus, yellow fever virus and the tick-borne flaviviruses, e.g., louping ill virus.

The family *Filoviridae*, a member of the order Mononegavirales, is the taxonomic home of several related filoviruses that form filamentous infectious viral particles (virions) and encode their genome in the form of single-stranded negative-sense RNA. Two members of the family that are commonly known as **Ebola** virus and *Marburg* virus. Both viruses, and some of their lesser known relatives, cause severe disease in humans and nonhuman primates in the form of viral hemorrhagic fevers

Ebola virus symptoms show up 2 to 21 days after infection and usually include: High fever, headache, joint and muscle aches, sore throat, weakness, stomach pain, and lack of appetite.

Ebola virus disease (EVD), formerly known as Ebola hemorrhagic fever, is a severe, often fatal illness in humans. Here are some facts as communicated by WHO[378]:

- The virus is transmitted to people from wild animals (fruit bats) and spreads in the human population through human-to-human transmission through body fluids.
- The average EVD case fatality rate is around 50%. Case fatality rates have varied from 25% to 90% in past outbreaks.
- The first EVD outbreaks occurred in remote villages in Central Africa, near tropical rainforests, but the most recent outbreak in West Africa (2014) has involved major urban as well as rural areas.
- Community engagement is key to successfully controlling outbreaks. Good outbreak control relies on applying a package of interventions, namely case management, infection prevention and control practices, surveillance and contact

tracing, a good laboratory service, safe and dignified burials and social mobilization.

It is thought that fruit bats of the Pteropodidae family are natural Ebola virus hosts, although they seem to be unaffected by it. Ebola is introduced into the human population through close contact with the blood, secretions, organs or other bodily fluids of infected animals such as fruit bats, chimpanzees, gorillas, monkeys, forest antelope or porcupines found ill or dead in the rainforest.

Ebola then spreads through human-to-human transmission via direct contact. Entry points for the virus include the nose, mouth, eyes, open wounds, cuts and abrasions. Contact with surfaces or objects contaminated by the virus, particularly needles and syringes, may also transmit the infection. While only being able to survive for a few hours on objects, the virus can survive for a few days within body fluids. The Ebola virus may be able to persist for three months (up to a full year!) in the semen, even after recovery, which makes sexual intercourse unsafe. Ebola may also occur in the breast milk of women after recovery, making it unsafe to breastfeed a baby.

EVD first appeared in 1976 in 2 simultaneous outbreaks, one in Nzara, Sudan, and the other in Yambuku, Democratic Republic of Congo (DRC). The latter occurred in a village near the Ebola River, from which the disease takes its name.

The 2014 outbreak in west Africa (first cases notified in March 2014) is the largest and most complex Ebola outbreak since the Ebola virus was first discovered in 1976. There have been more cases and deaths in this outbreak than all others combined. It has also spread between countries starting in Guinea then spreading across land borders to Sierra Leone and Liberia, by air (1 traveler only) to Nigeria, and by land (1 traveler) to Senegal.

The most severely affected countries, Guinea, Sierra Leone and Liberia, have very weak health systems, lacking human and infrastructural resources, having only recently emerged from long periods of conflict and instability. An experimental Ebola vaccine proved highly protective against EVD in a major trial in Guinea in 2015. The vaccine, called rVSV-ZEBOV, was studied in a trial involving 11,841 people. Among the 5837 people who received the vaccine, no Ebola cases were recorded 10 days or more after vaccination. In comparison, there were 23 cases 10 days or more after vaccination among those who did not receive the vaccine.

The rVSV-ZEBOV vaccine has been used in the 2018–2019 Ebola outbreak in DRC. It has been shown that the vaccine is highly effective.

However, WHO's Strategic Advisory Group of Experts has stated the need to assess additional Ebola vaccines.

A list of EVD risk reduction actions has been recommended by WHO. Above all, WHO advises to

- eliminate animal wildlife-to-human transmission by avoiding human contact with infected fruit bats, monkeys, apes, forest antelope or porcupines and the consumption of their raw meat. Animals should be handled with gloves and other appropriate protective clothing, and animal products (blood and meat) should be thoroughly cooked before consumption.

The *human immunodeficiency viruses* (HIV) include two species of *Lentivirus* (a subgroup of retrovirus) that infect humans. Over time they cause acquired immunodeficiency syndrome (AIDS), a condition in which progressive failure of the immune system—if not treated—eventually leads to death.

Both HIV-1 and HIV-2 are believed to have originated in non-human primates in West-central Africa, having transferred to humans ("*zoonosis*") in the early 20th century.

HIV-1 is thought to have jumped the species barrier on at least three separate occasions, giving rise to the three groups of the virus, M, N, and O.

HIV-1 appears to have originated in southern Cameroon through the evolution of SIVcpz, a simian immunodeficiency virus (SIV) that infects wild chimpanzees. The closest relative of HIV-2 is SIVsmm, a virus of the sooty mangabey (Cercocebus atys atys), an Old World monkey living in littoral West Africa, from southern Senegal to western Côte d'Ivoire. New World monkeys such as the owl monkey are resistant to HIV-1 infection, possibly because of a genomic fusion of two viral resistance genes.

HIV infects vital cells in the human immune system, such as helper T cells (specifically CD4+ T cells[379]), macrophages, and dendritic cells. HIV infection leads to low levels of CD4+ T cells through a number of mechanisms. In addition, infected CD4+ T cells are exposed to CD8+ cytotoxic lymphocytes that recognize and kill infected cells. When CD4+ T cell numbers decline below a critical level, cell-mediated immunity is lost, and the body becomes progressively more susceptible to infections, leading to the development of AIDS.

Fortunately, many years of intense biomedical research have now reduced the life-threatening aspects of HIV and turned AIDS into a chronic disease that can be managed.[380]

SARS, MERS, and COVID-19

Coronaviruses are widespread and can be found in most mammals. What distinguishes them and makes them such a threat to Sapiens is

   (i)   their ability to jump from animals to humans,

  (ii) their ability to readily exchange parts of their genetic material (RNA) with one another,

 (iii) their exceptional ability to confuse the human immune system,

 (iv) their ability to transmit from asymptomatic carriers.

During the past two decades, there have been three serious Coronavirus related outbreaks, namely SARS-1, MERS, and the SARS-CoV-2 pandemic.

SARS

Severe acute respiratory syndrome coronavirus (SARS-CoV or SARS-CoV-1)[381] is a strain of virus that causes infection of the epithelial cells within the lungs. The virus enters the host cell by binding to the ACE2 receptor.[382]

Angiotensin-converting enzyme 2 (ACE2) is attached to the outer surface (cell membranes) of cells in the lungs, arteries, heart, kidney, and intestines. By catalyzing the hydrolysis of *angiotensin II* (a vasoconstrictor peptide) into the vasodilator agent *angiotensin (1–7)*, ACE2 lowers blood pressure. ACE2 thereby counters the activity of the related angiotensin-converting enzyme (ACE) by reducing the amount of angiotensin-II and increasing Ang(1–7).[d] It is well known that several human blood pressure reducing drugs are "ACE inhibitors."[383]

---

[d]Angiotensin (1–7) is a vasodilator agent that plays important roles in cardiovascular organs.

As a transmembrane protein, ACE2 serves as the main entry point into cells for some coronaviruses. More specifically, the binding of the spike (S) protein[e] of SARS-CoV (and SARS-CoV-2, see below) to the enzymatic domain of ACE2 on the surface of cells results in endocytosis[f] and translocation of both the virus and the enzyme into endosomes[g] located within cells. This entry process also requires priming of the Spike protein by the host serine protease TMPRSS2,[h] the inhibition of which is under current investigation as a potential therapeutic

The human version of the ACE2 enzyme is often referred to as hACE2.

In the SARS outbreak of 2003, about 9–10% of patients with confirmed SARS-CoV infection died. The mortality rate was much higher for those over 60 years of age, with mortality rates approaching 50% for this subset of patients.

SARS infects humans, bats, and palm civets.

On 16 April 2003, following the outbreak of SARS in Asia and secondary cases elsewhere in the world, the World Health Organization (WHO) issued a press release stating that the coronavirus identified by a number of laboratories was the official cause of SARS. The Centers for Disease Control and Prevention (CDC) in the United States and National Microbiology Laboratory (NML) in Canada identified the SARS-CoV genome in April 2003. Scientists at Erasmus University in Rotterdam, the Netherlands, demonstrated that the SARS coronavirus fulfilled Koch's postulates, thereby confirming it as the causative agent. In the experiments, macaques infected with the virus developed the same symptoms as human SARS victims.

The first diagnostic SARS test was developed by Christian Drosten, a German virologist who has since become a leading global expert on Coronaviruses.

Phylogenetic analysis of these viruses indicated a high probability that SARS coronavirus originated in bats and spread to humans

---

[e]Spike proteins assemble into trimers on the virion surface to form the distinctive "corona" appearance.
[f]Endocytosis is a cellular process in which substances are brought into the cell.
[g]An endosome is a membrane-bound compartment inside a eukaryotic cell.
[h]Serine proteases are known to be involved in many physiological and pathological processes, e.g., are up-regulated in prostate cancer cells.

either directly or through animals held in Chinese markets. The bats did not show any visible signs of disease, but are the likely natural reservoirs of SARS-like coronaviruses. In late 2006, scientists from the Chinese Centre for Disease Control and Prevention of Hong Kong University and the Guangzhou Centre for Disease Control and Prevention established a genetic link between the SARS coronavirus appearing in civets and humans, confirming claims that the virus had jumped across species.

The spread of SARS-CoV could be contained although cases were reported 2002–2004 in many countries across Asia, and in Canada. WHO, then led effectively by Norwegian politician (and physician) Gro Harlem Brundtland,[384] is often credited for preventing SARS to become a serious public health problem outside Asia, by decisive actions such as enforced travel restrictions. In addition, the first SARS virus was not as infectious as SARS-CoV-2, to be covered below.

On 16 November 2002, the SARS outbreak began in China's Guangdong province, bordering Hong Kong.

In May 2005,[385] the New York Times wrote:

"Not a single case of the severe acute respiratory syndrome has been reported this year or in late 2004. It is the first winter without a case since the initial outbreak in late 2002. In addition, the epidemic strain of SARS that caused at least 774 deaths worldwide by June 2003 has not been seen outside of a laboratory since then."

MERS

Middle East respiratory syndrome (MERS), also known as camel flu, is a viral respiratory infection first discovered in 2012. It is caused by the MERS-coronavirus (MERS-CoV). Symptoms may range from none, to mild, to severe. Typical symptoms include fever, cough, diarrhea, and shortness of breath.[386] The disease is typically more severe in those with pre-existing health conditions.

MERS-CoV is a coronavirus believed to originally be transmitted from bats. However, humans are typically infected from camels, either during direct contact or indirectly. Spread between humans typically requires close contact with an infected person. Its spread is uncommon outside of hospitals. Thus, its risk to the global population is currently deemed to be fairly low.

In 2013, a wealthy 60-year-old Saudi Arabian man developed pneumonia, caused by a new virus that could be isolated from this patient. Its genetic material was first sequenced by the Erasmus Medical Center (EMC)[387] in Rotterdam, Netherlands, one of the world's leading virology institutes. Saudi Arabia then developed many similar cases and therefore had the greatest incentive to investigate the origin of the virus. Their Ministry of Health enlisted the help of Christian Drosten,[388] who since 2007 had been heading the Institute of Virology at the University of Bonn. His lab quickly developed a test to detect the genetic (RNA) component of the virus, and followed up with an antibody assay. Drosten's team found similarity with another well-known Corona virus (cold virus 229E discovered in 1966) that was known to have originated in camels. His collaboration with Marion Koopmans at EMC produced conclusive evidence that camels carry MERS-CoV.

In addition to the Saudi Arabian MERS outbreak in 2014, there were minor outbreaks in the Philippines (2014), in South Korea (2015) and in Kenya (early 2016). During the Kenyan outbreak, 500 camels died of MERS. No human fatality was reported but antibodies were found in healthy humans. In 2018, another Saudi Arabian outbreak occurred.

About 2500 cases have been reported as of January 2020, mostly in the Arabian Peninsula. The death rate is higher than the SARS death rate, at 35–40%.[389]

COVID-19

COVID-19 started in Wuhan, China, in December 2019, as the disease caused by SARS-CoV-2, a far more infectious form of the SARS coronavirus. Despite Chinese attempts to contain the virus to Wuhan by restricting domestic Chinese flights, the virus first escaped via international travel and has now infected a large part of the world. It reached the United States mostly via Europe and has made USA the leading country in numbers of infections and deaths. The World Health Organization declared the outbreak a "Public Health Emergency of International Concern" on January 30, and a "pandemic" on March 11, 2020.[390]

The virus is primarily spread via the respiratory route between people during close contact, most often via small droplets produced by coughing, sneezing, and talking, as illustrated via Tyndall effect[i] in Fig. 8.5.

Figure 8.5 Droplets generated by a person sneezing (source: James Gathany, CDC Library ID 11162).

A distance of 2 meters (~6 feet) between people is considered adequate to reduce infection risk in normal situations (people just talking, not having symptoms). The droplets do not travel through the air over long distances, they usually fall to the ground or onto surfaces. However, sneezing and coughing require a bigger distance than 2 meters, therefore, people who sneeze or cough are advised to isolate. Compared to SARS-1, this coronavirus is much more contagious and is therefore posing a much bigger threat to public health.

Less commonly than via direct respiration, people may also become infected by touching a contaminated surface and then touching their face. SARS-CoV-2 is most contagious during the first three days[391] after the onset of symptoms, although spread is possible before symptoms appear, and from people who do not show symptoms, thus violating Koch's initial (but later abandoned) first postulate, see Fig. 8.1.

---

[i]Scattering of light as a light beam passes through a colloid.

Coronavirus history and COVID-19 origin

There are several theories about COVID-19 "patient zero". The first known case may trace back to December 1, 2019 in Wuhan, Hubei Province, China. Over the next month, the number of coronavirus cases gradually increased. According to official Chinese sources, these were mostly linked to a seafood (and live animals) wholesale market. On December 27–28, the Sequencing company Vision Medicals[392] informed the Wuhan Central Hospital and the Chinese CDC of the results of a test related to a patient with respiratory disease symptoms. The disease was caused by a new coronavirus, later named SARS-CoV-2.

The Wuhan Municipal Health Commission made the first public announcement of a pneumonia outbreak of "unknown cause" on December 31, confirming 27 cases. In early and mid-January 2020, the virus spread to other Chinese provinces, helped by the Chinese New Year migration and Wuhan being a transport hub and major rail interchange. On January 20, China reported nearly 140 new cases in one day, including two people in Beijing and one in Shenzhen, the major Southern city close to Hongkong. A report in The Lancet, January 24,[393] indicated human transmission, strongly recommended personal protective equipment (PPE) for health workers, and said testing for the virus was essential due to its "pandemic potential." Details of this important LANCET paper are listed below:

- Chinese health authorities did an immediate investigation to characterize and control the disease, including isolation of people suspected to have the disease, monitoring of contacts, collection of epidemiological and clinical data, and investigation of possible diagnostic and treatment procedures.
- By January 7, 2020, Chinese scientists had isolated a novel coronavirus (CoV) from patients in Wuhan. The genetic sequence of the 2019 novel coronavirus (2019-nCoV) enabled the rapid development of point-of-care real-time RT-PCR diagnostic tests specific for 2019-nCoV. They are based on full genome sequence data that was placed on the Global Initiative on Sharing All Influenza Data [GISAID] platform on January 10.
- Cases of 2019-nCoV are no longer limited to Wuhan. Nine exported cases of 2019-nCoV infection have been reported

in Thailand, Japan, Korea, the USA, Vietnam, and Singapore to date, and further dissemination through air travel is likely.

• As of January 23, 2020, 835 confirmed cases were consecutively reported in 32 provinces, municipalities, and special administrative regions in China, including Hong Kong, Macau, and Taiwan. These cases and the recently documented infections in healthcare workers caring for patients with 2019-nCoV indicate human-to-human transmission and thus the risk of much wider spread of the disease. By January 23, 25 patients died.

Note that the term "2019-nCoV" is no longer in use; it has since been replaced by "SARS-CoV-2".

A map of Chinese provinces involved in the COVID-19 story is shown in Fig. 8.6.

Figure 8.6 From left to right, the figure shows the Chinese Hubei (Wuhan), Guangdong (Shenzhen), and Yunnan provinces.

The publication of the full genome sequence on GISAID[394] has enabled other tests, among them the official WHO test developed by the "Charité Global Health Center"[395] in Berlin, where Dr. Christian Drosten (see SARS section above) directs his research since 2017. Charité's diagnostic test protocol was completed within 3 days and then made available on WHO's website, allowing countries around the world to produce their own tests and therefore detect newly imported incidences of the new virus.

The animal origin of SARS-CoV-2 has been a subject of intense discussion. Based on the fact that the SARS virus originated with bats, the first hypothesis again pointed to bats. However, Wuhan is a city located in (cold) central China rather than the warm Southern Guangdong, and bats would rather be hibernating in caves instead of being very active. In addition, there is a significant difference between the genetic sequence of SARS-CoV-2 and hitherto known bat-related Coronaviruses. Also, SARS-2 is only about 76% identical to SARS-1.

A recent paper in *Current Biology* by Hong Zhou et al.[396] provides strong evidence that strengthens the hypothesis that SARS-CoV-2 is bat related and very similar to coronavirus RmYN02:

- Metagenomic analysis of 227 bats collected around Yunnan Province (se Fig. 8.5) between May and October, 2019, identified a novel coronavirus, RmYN02, from *R. malayanus*.
- RmYN02 was the closest relative of SARS-CoV-2 in most of the virus genome at 93.3% nucleotide identity. Parts of the virus that specify enzymes for replication show an even stronger similarity, 97.2%.
- However, two loop deletions in the ACE2 receptor binding domain (RBD) may reduce the binding of RmYN02, therefore reducing infectious risk.
- RmYN02 further contains an insertion at the S1/S2 cleavage site in the spike protein, providing strong evidence that such insertion events can occur naturally in animal betacoronaviruses.

The missing part of the SARS-CoV-2 puzzle may be that one of its parts closely resembles a pangolin coronavirus.

Pangolins (Fig. 8.7) are nocturnal anteaters that have large, protective keratin scales covering their skin; they are the only known mammals with this feature. They live in hollow trees or burrows, depending on the species.

Figure 8.7 Pangolin pup and its "curled up" mother (source: Wikipedia).

Pangolins are threatened by poaching. They are among the most trafficked mammals in the world. Humans are chasing them for their meat and scales. The scales are used in Chinese traditional medicine for a variety of ailments including excessive anxiety and hysterical crying in children, women accused of being possessed by devils and ogres, malarial fever, and even deafness.

We conclude therefore that—based on the habit of Coronaviruses to exchange parts with each other—it may be possible that SARS-CoV-2 includes elements from both bat and pangolin viruses. In any case, there seems to be little doubt that SARS-CoV-2 has animal origin.

COVID-19 testing

The history of COVID-19 testing is documented and updated regularly at Wikipedia.[397] Early developments were described in a paper by Cormac Sheridan in *Nature Biotechnology*.[398]

Most tests are based on use of the polymerase chain reaction (PCR), a method invented in 1983 and widely used in molecular biology to rapidly make millions to billions of copies of a specific DNA sample.

Other test methods such as immunoassays (antibodies), chest CT scans, and CRISPR-based paper strips are explained in the above-mentioned Wikipedia entry.

Note that PCR-based genetic tests are used to detect virus infections of patients. They look for the presence of the virus in the patient's body. On the other hand, "antibody tests" (or serological tests) are designed to detect previous infections and the associated reaction by the patient's immune system. Antibodies arise when the body is attacked by the virus.

Test samples for genetic tests are obtained from patients who have to either spit into a test tube or to provide a nasopharyngeal swab. Results are generally available within a few hours. There has been a controversy regarding saliva versus nasopharyngeal swab. A recent test conducted by the Yale University School of Public Health found that saliva yielded greater detection sensitivity and consistency throughout the course of infection when compared with samples taken with nasopharyngeal swabs that are getting less and less reliable as the virus moves from nose and throat to the lungs.

Antibody tests are based on blood samples and the results are generally returned faster. Such tests can also be scaled up to higher volumes than genetic tests, but are considered less accurate and less reliable.

Another option is to look for lung damage via either CT scan or low oxygen take up, even before the virus has established itself and can be reliably detected.

Since COVID-19 was first observed in China, the first test kits were developed there by BGI, China's leading Genomics company with global reach. BGI's PCR-based detection kit was approved by the Chinese equivalent of the FDA in January 2020. By the end of February, China had developed the manufacturing capability of more than a million test kits per day. Outside China, BGI's kit has received European CE-IVD certification.[j]

On January 19, it was announced that the well-known research hospital Charité in Berlin, Germany, had developed a PCR based test kit (see above), which subsequently formed the scientific basis for the official World Health Organization (WHO) COVID-19 test. Soon thereafter, 250,000 such kits were distributed to laboratories across the globe. A group at Hong Kong university developed another test (in two steps) that was based on previous Coronavirus tests and is also shared with WHO and over 30 labs in Asia, Africa, the Middle East, and South America.

A South Korean company called Kogenebiotech developed another high-quality test kit, named PowerChek Coronavirus by January 28, followed by two other Korean commercial test kits distributed in February. PowerChek looks for the "E" gene shared by all coronaviruses, and the "RdRp" gene specific to COVID-19.

In the US, the Centers for Disease Control (CDC) started distributing its PCR based diagnostic panel to public health labs in all 50 states as well as 30 international locations in early February, after having received "Emergency Use Authorization" by the US FDA. However, several state testing labs uncovered a quality issue with one element of the three-part assay, thereby seriously setting back the US testing effort. It is remarkable that USA, the acknowledged global medical research leader, was so late in rolling out a test, and then had to correct a serious quality issue, thereby delaying availability by several weeks.

---

[j]CE Marking is required for all in vitro diagnostic (IVD) devices legally sold in the EU (Europe) and indicates that an IVD device complies with the European in vitro Diagnostic Devices Directive (98/79/EC).

According to personal communication, Chinese scientists tried to help their US colleagues in meetings around February 7 to address the test kit issue. However, the FDA was reluctant to approve a non-US test kit. It's FDA's mission to protect the US public and make sure that drugs and devices are safe. In this particular instance, government bureaucracy may have been "overprotective" by not approving any of the available functioning international test kits, WHO/German, Korean and Chinese, all based on well-established scientific methodologies.

In the meantime, established commercial medical diagnostics leaders Roche Diagnostics,[399,400] Abbott Labs,[401] and Thermo Fisher,[402] have taken over and developed scalable COVID-19 tests.

The accuracy of a diagnostic test[403] is defined by its *sensitivity* (the correct identification of all patients with the disease). It is defined as the ratio of the proportion of the patients who have the condition of interest and whose test results are positive over the number who have the disease, *specificity* (the true negative rate), the proportion of people without the disease who will have a negative result. In other words, the specificity of a test refers to how well a test identifies patients who do not have a disease, *negative predictive value* (chance that a person with a negative test is truly disease-free), and *positive predictive value* (chance that a person with a positive test is truly diseased). Genetic tests typically have high analytical sensitivities and specificities under ideal circumstances. However, in clinical reality, the test sensitivity varies according to duration of illness (higher accuracy during early stages), site and quality of specimen collection, and the viral load. For COVID-19, an important factor is the patient's age. Younger people with strong immune systems are able to reduce the viral load in their bodies much quicker than patients 50 years or older. In older patients, viral loads can still be detected during the third week.

The accuracy of serological (antibody) tests can be near 100% when samples are acquired 20 days after infection or first symptoms. At earlier time points, the sensitivity and specificity are lower as the immune response is evolving. Inaccurate serological tests may lead to two major problems: the false labeling of patients who have been infected as disease-negative, and the false labeling of patients who have not been infected as disease-positive.

The Foundation for Innovative New Diagnostics (FIND), a Geneva-based not-for-profit organization, is collecting evaluations of COVID-19 molecular tests and immunoassays, in collaboration with the WHO and other partners.[404] The test situation (by country) has been updated daily at the Worldometer website.[405] The most relevant numbers are listed under the column heading "tests per million." From the beginning, there has been a strong correlation between the number of tests and the successful mitigation of the pandemic, by country. For example, South Korea and Germany were able to keep COVID-19 fatalities down by early testing, isolation of the infected, quarantining of contacts and restriction of group gatherings/events.

At home test kits are forthcoming.[406]

COVID-19 clinical picture

After over 8 million total global COVID-19 cases and at least 435,000 deaths (6–7% death rate but mostly confined to ages above 60) by mid-June 2020, the clinical picture is getting clearer. What is known is that the respiratory system is particularly affected. SARS-CoV-2 first enters the nose and throat, as shown in Fig. 8.8.

Tissues with a high presence of the ACE2 receptor, such as the lining of the nose, are highly vulnerable to infection. As the virus multiplies, an infected person sheds huge amounts of it, especially during the first week, even if the person is "asymptomatic." Typical symptoms are fever, dry cough, sore throat, loss of smell and taste,[k] and head and body aches, as indicated on Fig. 8.9. In some rare cases, patients also seem to develop diarrhea and nausea as initial symptoms.

If the infected person's immune system succeeds in beating back the virus as it attempts to travel down the windpipe, the patient may recover quickly. This seems to be the case for most young people, about 80% of all patients. However, if the virus attacks the lungs, it can turn deadly. As discussed in Chapter 4 (see Fig. 4.2), and shown again in Fig. 8.8, our thin respiratory branches end up in tiny air sacs, alveoli, that are each lined by a layer of cells rich in ACE2 receptors.

---

[k]As the virus is moving up the nose, it can damage cells in the nerve endings and affects the sense of smell.

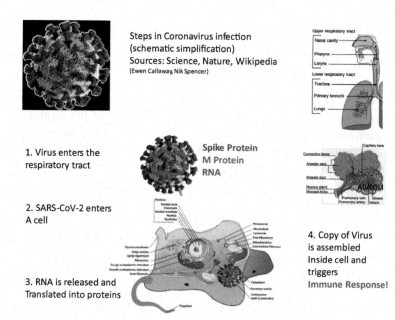

Figure 8.8 SARS-CoV-2 enters via the respiratory tract and uses the cell's machinery to replicate.

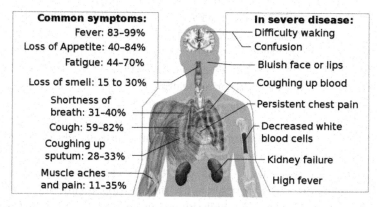

Figure 8.9 COVID-19 common and severe disease symptoms.

As explained in SCIENCE magazine,[407] the ensuing fight between the virus and the patient's immune system disrupts healthy oxygen transfer resulting in a severe drop of oxygen levels. CT scans of the lung reveal white opacities where normally black space (air) would be seen. The only chance for such patients will be to support their oxygen uptake with ventilators. Sadly, the ventilators often only delay the inevitable as the alveoli become stuffed with fluid, white blood cells, mucus, etc.

A driving force in the deterioration of seriously infected, and in particular male older patients, is an overreaction of the immune system known as "cytokine storm."[408] In Chapter 4, when discussing the Lymphatic System, we introduced the "cytokines," chemical signaling molecules that are part of sapiens' innate immune system. Viral infections may sometimes cause an overproduction of cytokines, and an associated attack on healthy tissues. A cytokine storm can cause blood vessels to leak (causing blood pressure to drop), blood clots to form and organs such as liver and kidneys to suffer life-threatening damage. This is something that is also known to happen during treatment by recent immune modulators for cancer.

Interestingly, there are cases of autoimmune initiation post-COVID-19. Among those cases, acute thyroiditis and type 1 diabetes are most common.[409]

SARS-CoV-2 is further affecting the cardiovascular system and the brain. The virus attacks not only the blood vessels but also the heart, in particular cardiac muscles. 20–40% of COVID-19 patients who are requiring intensive clinical care (ICU) suffer from heart damage, including reduced pump capacity and arrhythmia. Another COVID-19 side effect is blood vessel constriction, which is especially damaging in patients with prior health conditions such as diabetes mellitus, where the vessels are already affected.

A Dutch study[410] further revealed that the blood of about 40% of ICU patients clotted abnormally and that one third already had clots, leading to pulmonary embolism and stroke.

The fact that COVID-19 is targeting blood vessels provides an explanation why patients with pre-existing conditions such as diabetes, obesity and high blood pressure do face a higher risk. Another important risk factor is age: both our vascular and respiratory systems are getting weaker with age, and our immune response may also suffer.

In contrast, young people suffer much less and normally will not get seriously ill. However, New York, the principal COVID-19 hot spot in the US, has started an investigation into "childhood inflammatory disease" among children. Although only 1 percent of New Yorkers who have been hospitalized were under 20 years old, the NY State Department of Health is investigating[411] a number of cases (including 3 deaths) in New York children that have experienced symptoms (high fever and peeling skin) similar to Kawasaki disease[412] and toxic shock-like syndrome. 90% of those children have tested positive for COVID-19 either by genetic (diagnostic) or antibody testing.

In conclusion, COVID-19 is much more than a respiratory disease. The lung is the primary target of the virus but it also attacks the heart, the kidneys, the brain, and the vessels. Whenever there are multiple organ attacks, the patient will either die or suffer irreparable damage. As has been revealed by German data from an autopsy series of 27 patients,[413] SARS-CoV-2 can be detected in multiple organs, including the lungs, pharynx, heart, liver, brain, and kidneys.[414]

How epidemiologists hope to control and mitigate COVID-19

Epidemiology is a term derived from the Greek words *epi* (upon, among), *demos* (people), and *logos* (study, word, discourse), and defined as "study and analysis of the distribution (who, when, and where), patterns and determinants of health and disease conditions in defined populations."

When comparing epidemiology with molecular biology, it could be stated that epidemiology is a "social science" based on large numbers and statistical data about populations, whereas molecular biology is a physical science trying to arrive at a bottom-up understanding of the role of molecules, cells, tissues, and organs involved in health and disease of living organisms.

Epidemiologists leave it to molecular biologists to explain molecular mechanisms, the roles of genes and proteins, the mechanisms of action responsible for drug therapies, etc. Epidemiologists take a top-down view of what is going on when a virus such as SARS-CoV-2 causes a global epidemic. Based on their analysis of public health numbers, they model outcomes and then advise public health officials and governments on the most favorable course of action.

However, models are only as good as the assumptions used to develop them, and there will always be uncertainties in the choice of parameters and agreed goals. If the goal is to completely contain the infection of a population by a highly contagious virus such as SARS-CoV-2, the epidemiologist will recommend a complete "shutdown" of society, forcing everybody to stay at home, preventing people to meet and come close to each other, forbidding events involving more than 5 (or 10 or 50 or 100 participants), closing all schools and business activities such as travel, restaurants, sports events, etc.

The price paid by society for a complete shutdown is a very serious negative impact on economic growth, a significant jump in unemployment, and immense impact on mental health and social patterns.[415]

On the other hand, ignoring the virus and continuing activities like before would lead to huge outbreaks that could not be handled by the public health system, with consequences such as millions of people dying, at home or in the streets, and an even worse impact on world economy and trade.

Epidemiologists will therefore recommend a compromise between complete shutdown and complete neglect of the virus' existence, and they will further recommend ways to contain and mitigate the pandemic without sacrificing too many lives and without causing too much socio-economic and mental health damage. Their top-down approach will further benefit from close interaction with the bottom-up science activities performed by biomedical researchers.

Here is a list of tools available to epidemiologists and public health officials to optimize a country's response to COVID-19:

- *Hand washing*: hand hygiene is the act of cleaning one's hands to remove soil, grease, microorganisms, or other unwanted substances. Hand washing with soap—often, and throughout the day—prevents the spread of many diseases. People can also become infected with respiratory diseases, such as influenza or COVID-19, if they do not wash their hands before touching their eyes, nose, or mouth (i.e., mucous membranes).
- *Social distancing*: also called physical distancing, social distancing includes a set of interventions or measures taken to prevent the spread of a contagious disease by maintaining

a physical distance between people. For COVID-19, the recommended distance is 2 meters.

- *Quarantine*: restriction on the movement of people and goods, intended to prevent the spread of disease or pests. It is distinct from medical isolation, in which those confirmed to be infected with a communicable disease are isolated from the healthy population. For COVID-19, based on the known incubation time of 4–5 days, quarantine should last at least as long but can be lifted as soon as the quarantined person has been tested negative.

- *Covering coughs* and keeping unwashed hands away from the face.

- *Face coverings*: face masks and other "personal protective equipment" (PPE) are recommended for those who suspect they have the virus, and for their caregivers. During COVID-19, many public health organizations are requiring the use of face masks by the general public, whenever close contact cannot be avoided. The use of face masks has been controversial in some circles. However, a recent study published in the *LANCET*[416] should convince doubters that (in particular, well designed double-layer) face masks reduce infection risk, in particular in hospital environments. Most airlines have made use of face masks mandatory during flights, to protect their crews.

- *Contact tracing*: the process of identification of persons who may have come into contact with an infected person ("contacts") and the subsequent collection of information about these contacts. By tracing the contacts of infected individuals, testing them for infection, treating the infected and tracing their contacts in turn, public health authorities have traditionally reduced infections in the population. Contact tracing makes it possible to reduce the spread of an infection, to alert contacts to the possibility of infection and offer preventive counseling or prophylactic care. Contact tracing has been a pillar of communicable disease control in public health for decades and has been used to learn about the epidemiology of a disease in a particular population For COVID-19, countries who were early in implementing contact tracing along with testing (e.g., Singapore, South Korea,

Taiwan), were most successful in containing and mitigating the pandemic. Although funding and implementation remain challenging, the widespread use of smartphones should facilitate the future use of contact tracing. Both Apple and Google, who account for most of the world's mobile operating systems iOS and Android, respectively, are committed to COVID-19 tracking technology relying on Bluetooth Low Energy (BLE) wireless radio signals.

- *Screening* or focused testing: to answer questions like "what percentage of the population has been infected and has developed antibodies?" focused testing can be used. For COVID-19, where a large fraction of infected people may not suffer from any clinical symptoms, comprehensive testing can provide valuable information.

- *Herd immunity*: a form of indirect protection from infectious disease that occurs when a large percentage of a population has become immune to an infection, either through vaccination or previous infections that have generated antibodies. In a population in which a large proportion of individuals possess immunity, chains of infection are more likely to be disrupted. This stops or at least slows the spread of disease. For individuals that suffer from immunodeficiency, cancer patients on chemotherapy, or transplant patients that are taking drugs for immunosuppression, herd immunity is a crucial method of protection. Once a certain threshold has been reached, herd immunity gradually eliminates a disease from a population. Successful examples of herd immunity created via vaccination are smallpox (1977), and measles. However, the positive impact of herd immunity is threatened by opposition to vaccination in some countries.

It is interesting to compare the response to COVID-19 in a few representative countries.

China practiced rigorous shutdowns after the initial outbreak in Wuhan. As the country and the economy are opening up gradually, new outbreaks may happen. A recent shutdown[417] affecting 100 M Chinese just occurred to prevent a "second wave." A strong central government will do anything in its power to mitigate such occurrences. China's scientists were early in their attempts to develop SARS-CoV-2 tests and are trying hard to develop vaccines and treatments. China received praise for publishing the genome of the virus, but also criticism for letting the virus escape globally.

South Korea confirmed the first infected patient on January 20, 2020. A significant increase in COVID-19 cases was confirmed one month later because of religious travelers that visited Daegu from Wuhan: 2 days later, on February 22, 1,261 of 9,336 visitors to a church in Daegu reported symptoms. One day later, South Korea declared the highest level of alert, but on February 29, the number of confirmed cases had increased to 3150. South Korea introduced what is now considered the largest and best-organized response to screen the population for the virus, isolate any infected people, and trace and quarantine those who contacted them. Screening methods included mandatory self-reporting of symptoms by new international arrivals through mobile application. A key to S. Korea's success was the quick development and deployment of testing capability, to allow up to 20,000 people to be tested every day. S. Korea achieved containment and mitigation without quarantining entire cities. President Moon Jae-in's response to the crisis initially polarized the country but is now universally praised. However, fully re-opening the country remains a challenge: On May 15 it was reported that about two thousand businesses were told to close again when a cluster of a hundred infected individuals was discovered; contact tracing of 11,000 people was implemented to bring the situation under control.

Other countries that dealt well with the COVID-19 challenge are Singapore, Taiwan, and New Zealand. The numbers[418] tell a story of public health preparedness, timely testing, contact tracing and effective use of PPE's to protect healthcare professionals.

Italy confirmed the first case on January 31, when two Chinese tourists tested positive for SARS-CoV-2 in Rome. Cases then began to rise sharply, prompting the Italian government to suspend all flights to and from China and declare a state of emergency. An unassociated cluster of COVID-19 cases was later detected in Lombardy, starting with 16 confirmed cases. It took until February 22 for the Council of Ministers to announced a new decree-law to contain the outbreak. 50,000 people from eleven different municipalities in northern Italy were quarantined. On March 4, the Italian government ordered the full closure of all schools and universities nationwide as Italy reached a hundred deaths. All major sporting events, including Serie A football matches, were to be held behind closed doors until at least April. On March 9, all sports events were suspended completely for at least one month. On March 11, Prime Minister Conte ordered stoppage of nearly all commercial activity except supermarkets and pharmacies. However, the actions came too late. On March 19, Italy overtook China as the country with the most coronavirus-related deaths in the world after reporting 3,405 fatalities from the pandemic. The medical facilities were overwhelmed, in particular by elderly people who had been infected. The majority of the cases occurred in the rich Lombardy region with Milan as the capital. The hot spot was Bergamo, near Milan's international airport. Most fatalities were confined to men over 70 years of age and, thanks to serious lockdown measures, the respective "curves" of number of infections and number of deaths started to flatten towards the end of April.

Germany: The ongoing COVID-19 pandemic spread to Germany on January 27, 2020, when the first case was confirmed and contained near Munich, Bavaria. The majority of cases in January and early February originated from the same automobile-parts manufacturer as the first case. On February 25–26, multiple cases related to the Italian outbreak were detected further North in Baden-Württemberg. A large cluster linked to a carnival event was formed in Heinsberg, North Rhine-Westphalia, with the first death reported on March 9, 2020. New clusters were introduced in other regions via Heinsberg as well as via people arriving from China, Iran and Italy, from where non-Germans could arrive by plane until March 17–18. German disease and epidemic control is run by the Robert Koch Institute (RKI) according to a national pandemic plan. The German government and several health officials stated the country was well-prepared and did not initially implement special measures to stockpile medical supplies or limit public freedom. Since March 13, the pandemic has been managed in the "protection stage" as per the RKI plan, with German states mandating school and kindergarten closures, postponing academic semesters and prohibiting visits to nursing homes to protect the elderly. Two days later, borders to five neighboring countries were closed. By March 22, all regional governments had announced curfews or restrictions in public spaces. As of May 20, 2020, 178,494 cases had been reported with 8,263 deaths and approximately 155,700 recoveries. The surprisingly low preliminary fatality rate in Germany, compared to Italy, Spain and France, has been explained as follows: (i) higher number of tests performed, (ii) higher number of available intensive care beds with respiratory support, (iii) higher proportion of positive cases among younger people. Germany let epidemiologists be in charge and trusted virologist Christoph Drosten (see above) to be a consistent, reliable and truthful source of COVID-19 information. Germany's political leader, Angela Merkel, an accomplished scientist by training, made sure that Germany's response was driven by science and optimal under the circumstances.

Sweden differed from most other European countries in that it mostly remained open. Advised by state epidemiologist[l] Anders Tegnell who took over his position from his mentor Johan Giesecke[419] in 2015, the Swedish strategy was based on the following assumptions and principles: (i) Children are assumed to be unaffected by SARS-CoV-2, but people over 60, and in particular over 70, must be protected. (ii) Social Distancing and restrictions of large group meetings are necessary. (iii) Face masks are not required for the general public, only for healthcare workers. (iv) Disruption of business and associated consequences such as unemployment are almost as bad for public health as SARS-CoV-2 infections. (v) Air travel has to stop; no Swedes to leave and no international flights to arrive. (vi) Trust that the Swedish public will exercise caution, and trust that the Swedish welfare state and health care system will withstand the pressure.

Finally, Giesecke believes that his strategy will be sustainable and may even lead to (at least partial) herd immunity. Swedish daycare centers and schools (until age 14) have remained open, and even restaurants have continued to serve meals with fewer tables, no bar service, and social distancing practiced by customers. What did not go well, however, was the intended protection of homes for the old. Most COVID-19 deaths in Sweden occurred in nursing homes. The loose closedown strategy of Sweden is therefore resulting in a much higher number of deaths per million than in neighboring Nordic

---

[l]Per the Swedish Constitution, the Public Health Agency of Sweden has autonomy that prevents political interference and the agency's policy.

countries Norway, Denmark, and Finland, all with comparable social "welfare state" systems. However, Giesecke and Tegnell continue to claim that the Swedish strategy is optimal and will win in the long run, as the right balance of "saving lives and keeping the economy going." However, as death rates have continued to exceed Nordic neighbors, Tegnell and the Swedish government have admitted to "failure in protecting of the elderly." Poor adherence to PPEs and lack of discipline by nursing home staff have certainly played a role. The question is whether an increasing internal opposition, driven in particular by physicians and medical researchers, will convince a political majority to abandon the unique Swedish strategy.

As to other European countries, it is interesting that the previously much maligned *Greek* government did very well, that the small Nordic country *Iceland* set an example in mass testing thanks to scientific leader Kari Stefansson,[420] and that the UK switched from an initial "herd immunity strategy," similar to Sweden's approach, to a more conventional strategy that turned out to produce statistical infection and death toll numbers similar to Italy, *Spain*, and *France*. Despite initial outbreaks in some ski resorts, *Austria* is doing very well, and *Switzerland* is showing a perfect "flattening" of the curve after initial outbreaks close to the Italian border.

What about the *United States*, the country with the strongest medical research tradition: NIH, including the National Institute for Allergy and Infectious Diseases, led by legendary expert Anthony Fauci[421] since 1984; Centers for Disease Control and Prevention (CDC); Food and Drug Administration (FDA); and world leading bio-pharmaceutical industry with industrial giants J&J, Pfizer, Merck, BMS, Lilly, Amgen, Gilead, etc.?

On January 20, 2020, the first known case of COVID-19 was confirmed in the US Pacific Northwest state of Washington. "Patient zero" was a man who had returned from Wuhan on January 15. On January 31, the Trump administration declared a public health emergency and restricted entry for travelers from China who were not U.S. citizens. Unfortunately, the "White House" left an important door wide open for the virus to travel from China to USA indirectly, in particular via Europe to major US airports like New York. This door was left open for about 6 weeks, until mid-March 2020. In fact, SARS-CoV-2 analysis of US patients proved that specific virus mutations could be traced to Europe rather than China.

Another important mistake was made by the CDC who tried to develop their own testing kit, but experienced glitches in the rollout, as already covered above. The associated slow start in testing certainly obscured the extent of the outbreak. While the White House was in denial about the actual situation, FDA approval for foreign test kits (like Germany's WHO approved kit) was slow in being given, and criteria for people to qualify for a test were overly restrictive.

It then turned out that some States (like Washington and California) reacted faster than the federal US government, with better overall results, in particular mortality rates. As soon as the first death in the United States was reported in Washington state on February 29, Governor Jay Inslee declared a state of emergency. On March 6, President Trump signed the Coronavirus Preparedness and Response Supplemental Appropriations Act, which provided emergency funding for federal agencies to respond to the outbreak. Subsequently, Corporations encouraged employees to work from home, and Sports events and whole seasons were cancelled. Beginning on March 15, many businesses closed or reduced hours and schools across the country were shutting down. By March 17, the epidemic had been confirmed in all 50 states, and on March 26, the United States had more confirmed cases than any other country. Confirmed cases and deaths went up to exceed 25% of the respective total global numbers. New York had emerged as the COVID-19 epicenter with the highest death toll per capita. The whole country was caught unprepared, with "severe shortages" of test supplies, serious widespread shortages of PPE, and other strained resources due to extended patient stays while awaiting test results.

Despite warnings by the White House COVID-19 task force, led by Vice President Pence and including renowned experts such as Drs. Fauci and Birx, the federal government opened the door to "normalization," i.e. the loosening of restrictions, perhaps too early, and left it to individual governors to make local decisions. As a consequence, no consistent nationwide strategy was implemented. Some states, in particular states with Democratic governors, decided to be cautious and rather save lives than rushing to open up business as usual. Others, mostly governed by business-friendly Republican governors, were willing to take more risk and to tempt fate by disregarding alarming numbers that signaled increases in the number of infections and deaths. The US economy which is normally sustained about 70% by domestic consumption, has taken a tough hit, and unemployment has moved from a record low of less than 4% in January 2020 to a record high of ~15% in May 2020. Particularly hard hit are restaurants, the travel industry, and personal services such as gyms and barber shops. Schools will be closed at least until the fall. Another unforeseen consequence of the shutdown is the mental health implications and social patterns that have caused increased abuse of alcohol, drugs and domestic violence, especially among low-income families. The weak have been worst affected by the virus, as the elderly and the ones with prior health conditions are more vulnerable to morbidity and mortality, and the financially weak are more vulnerable to unemployment and lack of meals in schools and protection for their children.

As the US government experienced such unprecedented difficulties, it reexamined its relations with WHO and with China. On April 14, President Trump halted funding to the World Health Organization, accusing WHO to have mismanaged the pandemic. A few days later, he signed an executive order to temporarily suspend immigration to the United States. President Trump and Secretary of State Mike Pompeo further indicated that they would investigate the origin of the current pandemic, which by a very small minority of scientists has been tied to China's virology lab (rather than the fish and animal market) in Wuhan.

Finally, Dr. Anthony Fauci warned that a second wave of infections could easily hit the country—and the world—during the fall, as colder weather will be expected to facilitate a reemergence of COVID-19.

Are animals better protected against viral attacks, and what could we learn by studying animal models?

On April 13, 2020, Science Magazine published[422] an interesting summary of ongoing attempts to find animal models for the development of vaccines and treatments of COVID-19. So far, it has been found that

- **Hamsters** can be infected: An experiment with 8 hamsters at the University of Hong Kong has shown that the animals lost weight, became lethargic and started to breathe rapidly. Subsequent tests revealed that SARS-CoV-2 could be found in the hamsters' lungs and intestines. Both tissues are rich in

ACE2, the protein receptor described above. A comparison between ACE2 receptors in humans and hamsters shows that there are only 4 differences in the 29 amino acids of the critical domain where SARS viruses try to enter.

- *Monkeys* are used to perform preliminary tests of vaccines. Sinovac Biotech, a privately held Beijing-based company, tested two different doses of their COVID-19 vaccine with eight rhesus macaques. Three weeks later, the group introduced SARS-CoV-2 into the monkeys' lungs through tubes down their tracheas, and none developed a full-blown infection.[423] Monkeys are also studied at the Erasmus Univ. Medical Center in Rotterdam,[424] at Peking Union Medical College,[425] and at the Univ. of Wisconsin, Madison.

- *Mice* remain the preferred animal model for human disease. However, the regular ACE2 receptor in the mouse differs too much from the human version: in the critical domain of the ACE2 protein, the area where SARS-CoV-2 makes its entry, 11 of 29 amino acids differ from the human version. Therefore, regular mice are immune to the virus. The same is true for *rats* who show 13 differences. Is it still possible to breed mice that have their ACE2 receptors modified? In fact, S. Perlman et al, Univ. of Iowa, succeeded in doing that in 2007 to study SARS.[426] Since interest in SARS had decreased by then, Perlman gave the modified mouse to the mouse experts at Jackson Lab in Maine, the organization we already encountered in chapter 6A (pp 124–132). It is possible that "human ACE2 mice" supplied by JAX will enable breakthroughs in SARS-CoV-2 research. The scientists at JAX are also using CRISPR to change the sequence of the native mouse ACE2 gene, so that the corresponding protein is recognized by the virus. Another approach tried is to modify the virus such that it infects unaltered mice.

- *Rats* are also used as model organisms. In some experiments, their bigger size is an advantage, e.g. when repetitive bleeding is required. CRISPR can be used to create a rat model with a human ACE2 receptor. Rats may be better suited than mice to do vaccine studies and toxicology studies of drugs. It is ironic that 2020 is the Chinese "year of the rat". In Chinese old culture, Rat represents "the first one" or the beginning of a cycle of 60 years, and is said to represent wealth and surplus, as well as smartness, sensitive responses (escaping!), swift thinking, successful and strong surviving and reproductive ability. Let us hope that wealth and surplus will return and that humankind will manage to survive the COVID-19 crisis without major side effects such as depression or war!

- *Ferrets* are popular in influenza research because they experience symptoms similar to humans. Infected ferrets even sneeze! The SARS-CoV-2 virus does infect them but generally does not create severe symptoms, just an increase in body temperature. However, they demonstrated that even small respiratory droplets can drift in the air for long periods and over longer distances, thereby infecting ferrets in adjoining cages. What also has been observed with ferrets is that the virus strikes the elderly much harder: young ferrets had no symptoms while 93% of the older[427] died!

A recent Chinese study of domesticated animals[428] showed that SARS-CoV-2 replicates poorly in dogs, pigs, chickens, and ducks, but ferrets and cats are permissive to infection. The scientists, affiliated with Harbin Veterinary Research Institute, Chinese Academy of Agricultural Sciences, found experimentally that cats are susceptible to airborne infection.

How molecular biologists hope to develop vaccines and drug treatments

Disclaimer: What is covered and described below represents a snapshot in time (June 2020) and will certainly change as the global scientific community is working tirelessly to increase our understanding of COVID-19. We still believe that it is worth providing an overview of ongoing activities aimed at prevention and treatment of the infected.

The containment and mitigation measures discussed above are insufficient to return life to "normal," the way it was before COVID-19. However, a safe and effective vaccine will be needed to protect the human population against future outbreaks. Equally urgent is the search for a cure for COVID-19 patients, using the proven methods of drug discovery. Governments and philanthropic organizations around the world, as well as biopharmaceutical companies, are putting plans in place to reach those two important goals.

The status of COVID-19 vaccines under development is tracked by many organizations, among them the *New York Times*.[429] By June 17, 125 projects are still in pre-clinical phase, 8 have advanced to Phase I, another 8 have advanced to Phase II, and two current front runners, namely Oxford University, UK (with Astra-Zeneca), and the University of Melbourne, Australia (along with Mass. General Hospital, Boston, and UMC Utrecht).

Adaptive immune response:
1. SARS-CoV-2 virus is ingested by an antigen-presenting cell (APC). The virus is engulfed and T-helper cells are activated
2. T-helper cells (also known as CD4 cells) release T cell cytokines that suppress or regulate immune responses.
3. T cell cytokines activate B cell antibodies that can EITHER block the virus from further infecting cells, OR mark the infected cells for destruction by cytotoxic T cells.
4. Cytotoxic T (also known as CD8 or killer) cells identify and destroy virus-infected cells.
5. Memory and T cells recognize the virus and can monitor their attack on the body for a long time ("immunity").

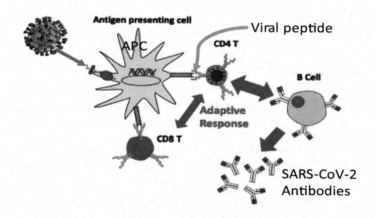

Figure 8.10   Vaccine basis: how we develop immunity.

The platforms based on DNA or messenger RNA offer considerable promise to alter COVID-19 antigen functions for strong immune responses, and can be rapidly assessed, refined for long-term stability, and prepared for large-scale production capacity. Other platforms being developed in 2020 focus on peptides, recombinant proteins, live attenuated viruses, and inactivated viruses.

To understand the various platforms, let us first review the steps involved when SARS-CoV-2 infects the human body, as presented by Callaway and Spencer in their "graphical guide explaining vaccine design."[430] Figure 8.10 shows the immune response which is exploited by vaccines. Note that, whenever a cell expresses foreign proteins, such as after viral infection, a fraction of these peptides will be displayed on the cell surface. Cytotoxic T cells specific for these peptides will recognize and kill such presenting cells.

Virus vaccines are based on using the virus itself, either in active (but weakened) form, or inactivated. Many existing vaccines (e.g. measles and polio) are developed this way. However, extensive safety testing is required.

Viral-vector vaccines are genetically engineered such that they can produce coronavirus proteins in the body. The "vector" is actually another virus (weakened that it cannot cause disease) that was manipulated in one of two ways: (1) it can still replicate within cells or (2) it cannot replicate because key genes have been disabled.

Nucleic acid vaccines are aiming to use genetic instructions (in the form of DNA or RNA) for a coronavirus protein. The goal is to generate an immune response that should help the human body to fight SARS-CoV-2. The given target protein is the spike protein. If successful, the immune response should result in antibodies that will make it difficult for SARS-CoV-2 to enter cells via ACE2.

Protein-based vaccines are based on the idea that injecting coronavirus proteins will generate antibodies that will protect the vaccinated person. Some projects are using fragment of proteins or "protein shells" that act like the coronavirus' outer coat.

Before moving into Phase I where vaccine safety is tested with volunteers, pre-clinical research is performed to assess vaccine safety and efficacy using COVID-19-specific animal models, such as ACE2-transgenic mice, other laboratory animals, and non-human primates. The handling of live viruses requires containment measures of high-level biosafety. Strong international coordination is required to ensure standardized safety procedures.

While the flu vaccine is typically mass-produced by injecting the virus into the eggs of chickens, this method will not work for the COVID-19 vaccine, as the SARS-CoV-2 virus cannot replicate inside eggs.

Another possible challenge will be to convince the currently 10% or more of the public that perceives vaccines as unsafe and unnecessary, for all kinds of reasons, religious and otherwise. Some people are expected to refuse vaccination and will therefore increase the risk of further COVID-19 outbreaks.

Among organizations that have formed international alliances to expedite vaccine development and prepare for distribution are

- WHO has implemented an access to COVID-19 tools accelerator for global vaccine development and has raised $8 billion.
- Coalition for Epidemic Preparedness Innovations (CEPI) is working with global health authorities and vaccine developers to raise funds in a global partnership between public, private, philanthropic, and civil society organizations for accelerated research and clinical testing of eight vaccine candidates. The CEPI goal is to support three candidates for full development to licensing. Early government participation has been received from UK, Canada, Belgium, Norway, Switzerland, Germany and the Netherlands.
- The Bill and Melinda Gates Foundation (BMGF) is (so far) donating US$250 million for research and public educational support, mainly in support of CEPI.
- Global Alliance for Vaccines and Immunization (GAVI) is financing and organizing clinical groups in under-developed countries with COVID-19 vaccination preparedness.
- the Global Research Collaboration for Infectious Disease Preparedness (GLoPID-R) works closely with WHO and member states to identify specific funding of research priorities needed for a COVID-19 vaccine.
- the International Severe Acute Respiratory and Emerging Infection Consortium (ISARIC) organizes and disseminates clinical COVID-19 information and will engage in eventual vaccine distribution.

Federal governments that are dedicating resources for national or international investments include

- *Canada*: Between March and late-April, the Canadian government announced funding for 96 research vaccine research projects at Canadian companies and universities, with plans to establish a "vaccine bank" of several new vaccines that could be used if another coronavirus outbreak occurs. On May 4, the Canadian government decided to contribute to WHO's goal to raise US$8 billion for COVID-19 vaccines and preparedness.
- *China*: The government is providing low-rate loans to vaccine developers through its central bank, and enabled land transfers to build production plants. There are nine Chinese COVID-19 vaccines in development. The government supports 3 Chinese vaccine companies and research institutes with special priority on efficacy. On May 18, China pledged US$2 billion to support WHO's overall efforts.
- *European Union*: CEPI's total investment in COVID-19 vaccine development has reached US$480 million by May. A research consortium includes Pasteur Institute, Themis Bioscience (Vienna, Austria), and the University of Pittsburgh. In addition, the European Commission provided an €80 million investment in CureVac (German Biotech) to develop a mRNA vaccine.
- *United Kingdom*: In April, the UK government formed a COVID-19 vaccine taskforce to stimulate British efforts for rapidly developing a vaccine through collaborations of industry, universities, and government agencies across the vaccine development pipeline, including for clinical trial placement at UK hospitals, regulations for approval, and eventual manufacturing. Funded initiatives are ongoing at the University of Oxford and Imperial College of London.
- *United States*: Biomedical Advanced Research and Development Authority (BARDA) a federal agency that funds disease-fighting technology, announced investments of nearly US$1 billion to support American COVID-19 vaccine development, and preparation for manufacturing the most promising candidates. BARDA invested in the

vaccine developer Moderna[431] and its partner, Johnson & Johnson. In addition, BARDA has a budget of US$ 4 billion to spend on vaccine development. The plan is to have a role in development of six to eight vaccine candidates to be in clinical studies over 2020–21 by companies, such as Sanofi Pasteur[432] and AstraZeneca.[433] On May 15, the US government announced federal funding for a fast-track program called Operation Warp Speed, which has the goals of placing eight diverse vaccine candidates in clinical trials by the fall of 2020, and manufacturing 300 million doses of a licensed vaccine by January 2021.

There has already been a lot of debate regarding possible cures for SARS-CoV-2. *Convalescent serum* lines up as the first-choice treatment,[434] and drugs like hydroxychloroquine,[435] remdesivir,[436] and dexamethasone[437] have been proposed and tested.

After previous positive results in some Chinese studies (Shenzhen's Third People's Hospital, and 3 hospitals in hot spot Wuhan), several leading medical research centers (such as Johns Hopkins, Mayo Clinic, Washington Univ, Einstein Medical Center and Mount Sinai-Icahn) have started a convalescent plasma program with US FDA, NIH, and industry partners.

A consortium[438] of GLOBAL blood plasma industry players including Japan's Takeda,[439] Australia's (with US HQ in King of Prussia) CSL Behring,[440] Biotest[441] (based in Germany and Boca Raton, but owned since 2018 by family business GRIFOLS with headquarters in Barcelona), UK's Bio Products Lab,[442] France's LFB,[443] and Switzerland's Octapharma.[444] The Bill & Melinda Gates Foundation is providing advisory support and Microsoft is providing technology including the Alliance website and the Plasma Bot for donor recruitment. The "CoVIg-19 Plasma Alliance" has developed a polyclonal antibody product, based on blood plasma of voluntary donors who have recovered from COVID-19. Donated plasma is pooled and sent to manufacturing facilities. There, the plasma is then processed to remove or inactivate viruses and concentrate instead on the antibodies. The result is called *hyperimmune globulin* (H-Ig). The CoVIg-19 Plasma Alliance will further work with the National Institute of Allergy and Infectious Diseases (NIAID) to test the safety and efficacy of the treatment in adults with COVID-19.

That study will lay the groundwork for a regulatory submission for the medicine.

In addition, BARDA is working with SAB Biotherapeutics and CSL Behring to introduce artificial chromosomes into cattle to produce human antibodies. Furthermore, GigaGen,[445] a San Francisco-based startup, is building a library of SARS-CoV-2 antibodies with the goal to introduce them into a mammalian cell line. GigaGen wants to produce "treatment cocktails" directed against the coronavirus causing COVID-19, in a scalable way that will not depend on the blood of individual donors.

A number of "Convalescent Plasma Trials" have been initiated in research hospitals around the globe, namely at

- Assistance Publique – Paris   120 patients
- Centre Hospitalier Ste Anne – Paris   138 patients
- China-Japan Hospital – Beijing   50 patients
- Erasmus Medical C – Rotterdam   426 patients
- Peking Union Medical C – Beijing   80 patients
- Puerta de Hierro Univ H – Madrid   278 patients
- Renmin Hospital – Wuhan   60 patients
- Nanchang Univ H – Nanchang   100 patients
- Wuhan Jinyintan H – Wuhan   100 patients

Under the leadership of immunologist Arturo Casadevall, Johns Hopkins University in Baltimore has spearheaded the use of a convalescent serum therapy in the US. FDA approval was given on April 8, 2020.[446]

Hydroxychloroquine, an old malaria drug, has been tested but could not been proven to help patients. Instead, it could trigger serious side effects such as heart arrythmia.

Remdesivir (developed by Gilead[447]) has shown a slight (but measurable) benefit. Gilead's antiviral research team will certainly try to build on the first positive (but also slightly controversial[448]) news.

When 2104 patients were given the low dose of 6 milligrams of dexamethasone for 10 days, and their outcomes were compared with those of 4321 patients receiving standard care, the steroid reduced deaths by one-third in patients already on ventilators and by one-fifth in patients receiving supplemental oxygen in other ways.[449]

However, the study did not find any benefit in patients not receiving respiratory support.

In addition to Gilead, Regeneron is also working with the US government: In addition to the COVID-19 vaccine and therapy projects already mentioned above, BARDA has chosen Regeneron as a partner in drug development.[450]

Societal and economic consequences

The COVID-19 epidemic has disrupted the global economy and global society in significant ways. Depending on personal value systems, several scenarios emerge as the desired outcome, as the preferred way to "get back to normal."

For believers in economic growth and never-ending prosperity based on capitalism and survival of the fit and strong, the vision of the future has not changed: going back to normal will be a return to exactly the way it was, only "better and stronger." There may be more careful preparation for the next pandemic, there may be some additional health checks (in addition to standard security) when entering office buildings or before boarding an airplane, but otherwise life will hardly differ from the way it was before.

Others may demand that the following questions need to be answered:

- What can we do to prevent viral diseases and future pandemics?
- Will there be (should there be) more or less global cooperation?
- Will there be a higher risk for confrontations between countries blaming each other for the recent crisis?
- Will there be more respect for our planet's environment we all live in and depend on?
- Will there be more respect for and better understanding of other species than sapiens?

Certainly, viruses are not popular after having just caused so much trouble for humanity. However, as pointed out by Petrov et al. (Stanford University Biology Department),[451] viruses have played and are continuing to play an important role in driving evolution of all living species on our planet. They are estimating that 30% of all protein adaptations in human evolution (after our separation from chimpanzees) were due to viruses. According to Prof. Petrov, humans are in no small ways shaped by our constant and ongoing battle with viruses.

Recent outbreaks caused by influenza, Ebola, and coronaviruses could be locally contained, but SARS-CoV-2 was more infectious (and more devious since non-symptomatic patients spread the virus!) and caused a pandemic. Another virus, HIV, has also caused a lot of harm, before a significant medical research effort finally paid off, reducing AIDS to a chronic disease that can be managed.

What further characterizes the current COVID-19 pandemic is its zoonotic nature, the animal origins of the virus. It is not only the careless nature of Chinese wild animal markets, there is also an element of "losses in wildlife habitat quality and biodiversity," as pointed out by scientific studies[452] that are critical of our focus on economic growth, fossil fuel consumption, human population growth, and disrespect for the environment. Nathalie Seddon et al therefore propose to incorporate the values of biodiversity into decision-making using economic methods, and develop several lines of argument for how biodiversity might be valued, building on recent developments in natural science, economics and science policy processes.

In a thoughtful essay published in *Yes Magazine* and titled "The Light at the End,"[453] Nafeez Ahmed is advocating that it is time for humanity to "imagine the world anew." In his view, when we will emerge from COVID-19, there will be no simple "normal" to which to return. Instead of clinging to the past, we should have the courage to embrace a different paradigm that respects our planet and the living organisms that inhabit the Earth.

Here are a few preliminary ideas, extracted from his essay:

We will need to invest in forms of energy and manufacturing less prone to depletion, and demand sustainability and resilience of households, communities, and businesses. We need to stop our addiction to shopping, our hunger for more and more "stuff" filling our apartments and houses. We need to make agriculture more sustainable by reducing pesticides and fertilizers and avoiding soil degradation. Humans in decision-making positions need to steer society away from material accumulation and greed.

Is the author naïve to think that a sustainable and compassionate world can be created?

Is it really possible for countries to reduce nationalistic feelings and the strong desire to "win" at any price, instead of working together and solving the bigger problems that current and possibly future pandemics - as well as climate change are creating for planet Earth?

COVID-19 is perhaps only a first warning of things to come. Can we really afford to develop protectionist behavior and isolate rather than collaborate and help each other?

The global science community is showing that collaboration across the world is possible. When searching for new insights into our understanding of life, political differences and national boundaries are irrelevant. There is no "national science," the quest for knowledge is universal and it connects scientists around the world. And knowledge should be shared because meeting our current and future challenges will require our combined efforts.

Conclusion

From ancient times to the present, humans and animals have lived together and shared the planet Earth. We interact constantly and we depend on each other. Animals and microbes contribute abundant material resources to humans and provide infinite benefits. Through observation and research, humans have studied animals and been impressed by their capabilities and their amazing adaptations to the particular environment they live in.

The more we learn, the more we are humbled by the diversity of life on our planet and by the realization that we are not as special as we may think. We have a lot in common with our relatives, we share many basic biological mechanisms that enable life on Earth, but we have certainly evolved our brain a bit further. What also separates sapiens from any other living species is the ability to compensate for our weaknesses and disadvantages: when we realize that we cannot fly we learn to build airplanes, when our vision cannot compete with "best practices" among animals, we develop binoculars and microscopes, and the list goes on.

What we are emphasizing in this book is that even our ambition to maintain and improve our human health can benefit from the study of animals. We are presenting examples where particular animal connections provide existing proven benefits, as well as potential future benefits for the enhancement of scientific knowledge, human health, and performance.

We are excited and surprised by what we have found, and convinced that our current compilation of animal stories represents just the beginning of what our companions can teach us. We are convinced that we have just scratched the surface and that our collection of examples is rather incomplete. There is much more we can learn from a closer connection to species other than our own. We hope that this book will stimulate such efforts and perhaps even triggers extra funding.

We hope that this book will lead to more interdisciplinary contacts between biologists, geneticists, zoologists, and medical researchers, as well as veterinary and human medical doctors. We hope that the breakthroughs described in this book will translate into benefits not only for human patients, but for our planet's ecosystem. The authors of this book are also strongly endorsing bioethical guidelines in all animal-related human activities, including life sciences research projects. Wherever possible, we propose to carry out genomics and other -omics studies, cell and tissue biology experiments, genetic engineering, and computer simulations rather than animal experiments. Out of respect for our animal friends, we hope that future life sciences research and biopharmaceutical R&D will rely more and more on in vitro and in silico experiments instead of in vivo animal studies.

Humans dedicating their lives to the study of animals seem to have one thing in common, namely love and respect for the objects of their studies. Humans in diverse cultures around the world have started movements and activities that reflect these ideas. Gautama Buddha is cited as saying that "when a man has pity on all living creatures then only is he noble," and various Western philosophers and writers have expressed strong feelings for animals.

The German philosopher Friedrich Nietzsche has written that "the animal has its right like man, so let it run about freely; and you, my dear fellow man, are still this animal, in spite of all!"

The 19th-century French naturalist and popular author of books on the lives of insects, Jean-Henri Fabre, has been called by Charles Darwin "an inimitable observer" of insects who combined a passion for scientific truth with an engaging, colloquial style of writing.

The English-born author and wildlife artist Ernest Thompson Seton became one of the founding pioneers of the Boy Scouts of America and an early pioneer of animal fiction writing.

The British writer Rudyard Kipling wrote *The Jungle Book* and the American author Jack London wrote *The Call of the Wild* in the same naturalistic spirit of respect for animals in the wild.

Ernest Hemingway, who is often accused of his love of bull-fights and for glorifying the killing of animals, later in life developed sympathy and compassion for animals. In "An African Story," a boy wanted to "help the elephant," and in the *Green Hills of Africa,* the hunter who was hospitalized because of a broken right shoulder, thought that "maybe what I was experiencing was a punishment for all hunters." Hemingway ended up respecting (although not openly advocating) the idea of "equality of all living beings."

We should accept animals for what they are and consider humans as equals, not as the masters that can do whatever they want, without consequences for life on our planet. Knowing animals and understanding animals means knowing ourselves and understanding ourselves.

Considering the discoveries yet to be made in the animal world, we have a responsibility as humans to preserve and enhance natural ecosystems rather than destroying them. One of the great challenges of the future will be how to live in symbiosis with animals, leveraging their strengths for their own and our future quality of life.

Appendix

Metric or US "Customary Units" System?

Before landing at Stockholm airport, the Scandinavian Airlines pilot makes a special announcement: "We are now leaving our cruising altitude of 33,000 feet—or 10,000 meters—and will land in about 30 minutes at Arlanda. The current temperature on the ground is 7 degrees Celsius, which is about 45 degrees Fahrenheit."

Why is the pilot communicating two sets of equivalent data for distance and temperature? The answer is that a small fraction of the world's population—about 5%—is still resisting the conversion to the metric system. It also happens to be the case that the tiny 5% represent the most powerful country on Earth.

The United States of America have resisted going "metric" while even member states of the conservative British Commonwealth decided to do make this important change by mid-20th century: India switched in 1954, the U.K. in 1965, and Australia and New Zealand in 1969. Even Canada and Mexico, both having extensive trade relations and sharing thousands of miles of borders with the USA, have adopted the metric system.

The metric units[364] are easy to use because they are based on multiples of 10, thereby enabling very easy multiplication and division by 10, 100, 1000, etc., just by adding zeroes, or by moving the decimal point in a real number. The metric system was introduced after the French Revolution and was initially based on just length and weight. The basic length unit was called the "metre" and the unit for weight was named the "gramme." In 1866 the United States started to use the metric system, but more than 150 years later it still is not the universally accepted main US system of measurement. By 1875, many countries in Europe and in Latin America had changed to using the metric system. A new organization called the International

Bureau of Weights and Measures (BIPM) was set up. In 1960 the rules for the metric system were revised to the International System of Units (often abbreviated SI). The SI definitions also include rules for abbreviating SI quantities across all languages. In the 1970s many people in the United Kingdom and the rest of the Commonwealth started using the metric system at their places of work.

In 1866, the United States passed a law that allowed people to use either the metric system or United States *customary units* for trade. In 1893 the United States Congress passed the Mendenhall Order which defined the yard as being exactly 3600/3937 metres and the pound as being exactly 0.4535924277 kilogram. The order changed the definitions (by accepting "primary" metric units) but had no other effect on people's lives.

In 1975 the Metric Conversion Act started a formal (but still voluntary) metrication process, coordinated by the U.S. Metric Board. In 1988 the congressional Trade and Competitiveness Act stated that metric units had to be used for all federal projects. However, the law did not apply to state projects. The resulting chaotic situation was that some states used metric units while other states did not. Also, some industries changed to using metric units but others did not. Soft drinks are sold in metric quantities while milk is still packaged in customary units. Metric units are widely used in the design of automobiles while advanced aircraft such as the Boeing 787 Dreamliner were designed using mainly customary units.

Public opinion is still divided about completion of the change-over to the metric system. Most experts agree that metrication can only work if all fifty states take the final step at the same time, as coordinated by the Federal Government.

Having waited so long, it is probably too late for the US to give up the current hybrid system that still relies on customary units and traditional names. Although there is no official government number sizing the cost versus the benefits, there seems to be agreement that going "metric" would be too expensive. For just one US government agency, NASA, the estimated expense would be above $300 million. However, not converting also carries a cost. For instance, NASA may have lost $125 million when its Mars Climate Orbiter was destroyed after its altitude-control system mixed up U.S. customary units with metric units.[365] Other industries could face different conversion costs, although U.S. exports would probably benefit in the long run.

When discussing both air and human body temperature, many US citizens feel that their traditional system of measurement remains a "pillar of American individualism." That's a strange thing to claim given the history of temperature scales: What's so "American" about Daniel Gabriel Fahrenheit[366] (1686–1736), who was born into a German family based in Danzig (now Polish territory and renamed Gdansk), a Hanseatic city? He was a gifted physicist, inventor and scientific instrument maker and spent his most productive years (1701–36) in the Dutch Republic as one of the figures arriving towards the end of the Golden Age (1590–1720) of Dutch science, technology, and the arts. Notables associated with the 17th-century Dutch Golden Age were (among others) Descartes, Huygens, Rembrandt, and Vermeer. A contemporary (to Fahrenheit) Dutch scientist was Antonie Philips van Leeuwenhoek (1632–1723), who is famous for having been one of the first microscopists and microbiologists. Fahrenheit (see Fig. A.1) invented the mercury-in-glass thermometer and introduced the first standardized temperature scale to be widely used. Since the early 1710s until the beginnings of the electronic era, mercury-in-glass thermometers were among the most reliable and accurate thermometers ever invented.

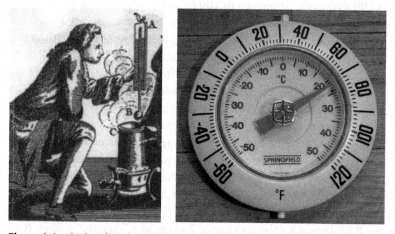

Figure A.1   D. G. Fahrenheit and his temperature scale (source: Wikipedia).

Based on his scientific achievements, Fahrenheit deserves our respect and admiration. However, when the Swedish scientist Anders Celsius proposed his "metric" (or "centigrade") temperature scale in 1742, there was less reason to keep the far more complicated Fahrenheit scale. In particular, when the metric system was introduced in 1791, it became obvious that the Celsius scale was perfectly compatible and a much better fit.

Here is how the competing scales are defined:

- Zero degrees Fahrenheit corresponds to about MINUS 17.8 degrees Celsius. Fahrenheit defined this temperature as the equilibrium of ice, water, and a salt ("ammonium chloride or even sea salt").
- Zero degrees C (Celsius) corresponds to 32 degrees F (Fahrenheit) and is defined by the freezing of pure liquid water (without salt) to turn into solid ice.
- Human body temperature is about 36.5 degrees C or 97.7 degrees F.
- One hundred degrees C corresponds to 212 degrees F and is defined by boiling water at sea level.
- The absolute zero point (Zero on the Kelvin[367] scale) is –273.15 degrees Celsius or –459.67 degrees F.

The temperature range between freezing and boiling pure water is 100 degrees C compared to 180 (= 212 – 32) degrees F.

Although the process of converting temperatures from F to C (and back) is a nice way to keep the brain engaged in some easy arithmetic, it remains a challenge to do it on a regular basis. To go from F to C, you have to subtract 32 and then multiply with 5/9. To go from C to F, you first multiply by 9/5 and then add 32. The scales coincide in only one point: –40 degrees F = –40 degrees C.

Aside from temperature, the measurement of volumes may have been the most confusing of all global agreements on various units. There are 16 US fluid ounces in a US pint but there used to be 20 imperial [UK] fluid ounces in an imperial pint. The US fluid ounce is larger than the now obsolete imperial fluid ounce, but the imperial pint was larger than the (still existing) US pint. In addition, there are 8 pints in a gallon and hence 2 pints in a "quarter" gallon. It is therefore not always easy to convert pints to liters, even for beer lovers! With wine bottles, on the other hand, the metric system seems to have conquered even US wine producers: A standard bottle holds 0.75 liters and nobody is trying to convert that to pints or gallons (although it would be close to a "fifth").

Having stated some historic facts and anecdotal stories, let's now move to some detailed and hopefully useful "metric—US customary" conversion numbers:

Distance

- 1 US mile is 1609 meters or 1.609 kilometers. A quarter mile is a little more than 400 meters
- 1 inch is 2.54 centimeters. 12 inches = 1 foot or 30.48 cm. 3 feet = 1 yard or 91.44 cm
- 1 kilometer = 1000 meters = 10^3 meters
- 1 centimeter = 0.01 meter = 10^{-2} meter
- 1 micrometer = 10^{-6} meter
- 1 nanometer = 10^{-9} meter
- When trimming a beard, a "stubble look" will result if the trimmer is set to 3/64 inches (1 mm)
- However, there is no recommended use of "nano-inches" as units of length in the molecular world

Velocity

- 1 mph = 1.609 kmh; 60 mph = 96.54 km per hour
- 10 meters per second = 3600 × 10 meters per hour = 36 kmh
- Velocity of light c = 300,000 km per second, or 3×10^{10} cm per second

Weight

- 1 pound (lb) ≈ 0.45359 kg; 100 kg ≈ 220 lb
- 10^3 kg = 1000 kg = 1 ton; 10^{-3} kg = 1 gram = 0.001 kg
- 1 lb = 16 ounces (oz)

Volume

- 1 liter (l) = 33.8 oz, 1 ounce (oz) = 29.58 ml = 0.02958 l
- 1 pint (pt) = 16 oz = 473 ml
- 1 gallon = 64 oz = 8 pt = 8 × 0.473 l = 3.784 l
- 4 oz = 118.3 ml, 6 oz = 177.3 ml, 8 oz = 237 ml

- 1 quart = 32 oz = 946.7 ml
- 100 ml = 3.38 oz
- 1000 ml = 1 l = 33.8 oz

One of the most confusing things when moving between the USA and ROW (rest of world) is how to measure gasoline consumption of automobiles: In the US it's "miles per gallon," whereas ROW uses "liters per 100 km." That's a tricky conversion to do in your head! For example, 30 mpg = 30 × 1.609/3.784 = 12.756 km/l ≈ 7.84 l/100 km.

The bottom line is that people will prefer the units they grew up with, and that change is difficult and probably hard to accept. Unless converted at an early age, people will be reluctant to give up their customary units.

Actually, even the ROW users of the metric system are guilty of a universal violation, namely by the persistent use of *horsepower* when measuring the performance of their cars. Given that horsepower is also part of the US Customary Unit system, we seem to have found the one unit that unifies the world–at least until the expected future dominance of electric cars!

As long as airline pilots are willing to translate temperatures, distances, etc. between metric and US customary measurement units during their announcements, the world will not change.

References

1. Hehenberger, M (2015). *Nanomedicine: Science, Business, and Impact,* www.panstanford.com

2. http://www.nobelprize.org/nobel_prizes/chemistry/laureates/1980/berg-lecture.html

3. Schrödinger, E (1944). *What Is Life?: The Physical Aspect of the Living Cell.* Based on lectures delivered under the auspices of the Dublin Institute for Advanced Studies at Trinity College, Dublin, in February 1943.

4. Watson, JD, Crick, FHC (1953). Molecular structure of nucleic acids: a structure for deoxyribose nucleic acid. *Nature,* **171**(4356): 737–738.

5. http://www.nobelprize.org/nobel_prizes/medicine/laureates/1968/

6. http://www.nobelprize.org/nobel_prizes/medicine/laureates/1962/crick-lecture.html

7. http://www.nobelprize.org/nobel_prizes/medicine/laureates/1962/

8. http://www.nobelprize.org/nobel_prizes/chemistry/laureates/1993/mullis-lecture.html

9. http://www.nobelprize.org/nobel_prizes/chemistry/laureates/2009/yonath_lecture.pdf

10. http://www.nobelprize.org/nobel_prizes/chemistry/laureates/1962/perutz-facts.html

11. Spang, A, Saw, JH, Jørgensen, SL, Zaremba-Niedzwiedzka, K, Martijn, J, Lind, AE, van Eijk, R, Schleper, C, Guy, L, Ettema, TJG (2015). Complex archaea that bridge the gap between prokaryotes and eukaryotes. *Nature,* **521**: 173–179.

12. https://www.nobelprize.org/prizes/medicine/1965/summary/

13. http://www.nobelprize.org/nobel_prizes/chemistry/laureates/1988/

14. http://www.nobelprize.org/nobel_prizes/chemistry/laureates/1957/

15. http://www.nobelprize.org/nobel_prizes/chemistry/laureates/1978/mitchell-lecture.pdf

16. http://www.nobelprize.org/nobel_prizes/chemistry/laureates/1997/boyer-lecture.pdf

17. Boyer died in 2018, 100 years old. In his obituary, his former student Eisenberg, DS shared interesting thoughts about Boyer's engineering talents that enabled him to solve the ATP problem: http://science.sciencemag.org/content/361/6400/334

18. http://www.nobelprize.org/nobel_prizes/chemistry/laureates/1997/walker-lecture.html

19. http://www.nobelprize.org/nobel_prizes/chemistry/laureates/1997/skou-lecture.html

20. http://www.nobelprize.org/nobel_prizes/medicine/laureates/1974/duve-facts.html

21. http://www.nobelprize.org/nobel_prizes/medicine/laureates/1946/muller-facts.html

22. Doudna, JA, Charpentier, E (2014). The new frontier of genome engineering with CRISPR-Cas9. *Science*, **346**(6213): 1258096.

23. http://www.sciencemag.org/news/2017/02/biologists-propose-sequence-dna-all-life-earth

24. Tudge, C (2000). *The Variety of Life*. Oxford University Press. ISBN 0198604262.

25. Rouse, GW (2002). *Annelida (Segmented Worms)*, Encyclopedia of Life Sciences. John Wiley & Sons, Ltd. doi:10.1038/npg.els.0001599.

26. Laybourne, RC (1974). Collision between a vulture and an aircraft at an altitude of 37,000 feet. *Wilson Bull*, **86**(4): 461–462. ISSN 0043-5643.

27. Mouritsen, H, Ritz, T (2005). Magnetoreception and its use in bird navigation. *Curr Opin Neurobiol*, **15**: 406-414.

28. https://bigthink.com/philip-perry/the-mystery-of-how-birds-navigate-is-over-and-the-answer-is-so-amazing

29. Berta, A, Sumich, JL (1999). *Marine Mammals: Evolutionary Biology*. Academic Press, San Diego. ISBN 978-0-12-093225-2. OCLC 42467530.

30. https://en.wikipedia.org/wiki/Fish

31. https://evolution.berkeley.edu/evolibrary/article/fishtree_02

32. Dorit, RL, Walker, WF, Barnes, RD (1991). *Zoology*. Saunders College Publishing.

33. http://www.sciencemag.org/news/2016/08/greenland-shark-may-live-400-years-smashing-longevity-record

34. Foley, LE, Gegear, RJ, Reppert, SM (2011). Human cryptochrome exhibits light-dependent magnetosensitivity. *Nat Commun*, **2**: Article no. 356.

35. Andrews, RD, et al. (1997). Heart rates of northern elephant seals diving at sea and resting on the beach. *J Exp Biol*, **200**(Pt 15): 2083–2095.

36. Gibson, RN, Atkinson, RJA (2007). *Oceanography and Marine Biology: An Annual Review*, Vol. 41. CRC Press.

37. Hulley, PA (1998). Paxton, JR, Eschmeyer, WN (eds.). *Encyclopedia of Fishes*. Academic Press, San Diego, pp. 127–128.

38. Linnaeus, C (1735). *Systema Naturae*.

39. de Monet de Lamarck, J-BPA (1809). *Philosophie zoologique: ou Exposition des considérations relative l'histoire naturelle*, Vol. 1.

40. Hopwood, N (2015). *Haeckel's Embryos: Images, Evolution, and Fraud*.

41. http://kenstoreylab.com/wp-content/uploads/2016/03/Biochemical-adaptation-to-extreme-environments.pdf. From: Walz, W (ed.), *Integrative Physiology in the Proteomics and Post-Genomics Age*. Humana Press Inc., Totowa, NJ.

42. Kelley, JL, et al. (2014). Compact genome of the Antarctic midge is likely an adaptation to an extreme environment. *Nat Commun*, **5**: 4611.

43. Yu, L, et al. (2016). Genomic analysis of snub-nosed monkeys (Rhinopithecus) identifies genes and processes related to high-altitude adaptation. *Nat Genet*, **48**(8): 947–952.

44. Zhang, Z, et al. (2016). Convergent evolution of rumen microbiomes in high-altitude mammals. *Curr Biol*, **26**(14): 1873–1879.

45. Pääbo, S (2015). *Neanderthal Man: In Search of Lost Genomes*.

46. Yang, JL, Sykora, P, Wilson, DM, Mattson, MP, Bohr, VA (2011). The excitatory neurotransmitter glutamate stimulates DNA repair to increase neuronal resiliency. *Mech Ageing Dev*, **132**(8–9): 405–411.

47. Eagleman, D (2015). *The Brain: The Story of You*. www.vintagebooks.com

48. Kandel, ER (2016). *Reductionism in Art and Brain Science*. Columbia University Press.

49. Lacey, S, Sathian, K (2012). Representation of object form in vision and touch, In *The Neural Basis of Multisensory Processes*, Murray, MM, Wallace, MT, (eds.), Chapter 1. CRC Press, Boca Raton, FL.

50. https://www.nobelprize.org/nobel_prizes/medicine/laureates/2004/

51. https://www.nobelprize.org/nobel_prizes/medicine/laureates/2014/

52. https://www.iasp-pain.org/Taxonomy

53. A big topic, discussed extensively by many deep thinkers. An interesting point of view was presented by Stephen Asma in the New York Times on June 3, 2018.

54. Dawkins, R (2006). *The God Delusion*. Houghton Mifflin, Boston, p. 406. ISBN 0-618-68000-4.

55. Collins, FS (2006). *The Language of God: A Scientist Presents Evidence for Belief.* Simon & Schuster, ISBN 0-7432-8639-1.

56. Kurzweil, R (2005). *The Singularity is Near*. Viking Books, New York. ISBN 978-0-670-03384-3.

57. World Health Organization (WHO) (2015). http://www.who.int/gho/mortality_burden_disease/causes_death/top_10/en/

58. https://www.cdc.gov/nchs/fastats/deaths.htm

59. Dumas, L, et al. (2007). Gene copy number variation spanning 60 million years of human and primate evolution. *Genome Res*, **17**(9): 1266–1277.

60. Varki, N et al. https://www.ncbi.nlm.nih.gov/pmc/articles/PMC3352420/

61. Schreiweis, C, et al. (2014). Humanized Foxp2 accelerates learning by enhancing transitions from declarative to procedural performance. *Proc Natl Acad Sci U S A*, **111**(39): 14253–14258.

62. Popesco, MC, Maclaren, EJ, Hopkins, J, Dumas, L, Cox, M, Meltesen, L, McGavran, L, Wyckoff, GJ, Sikela, JM (2006). Human lineage-specific amplification, selection, and neuronal expression of DUF1220 domains. *Science*, **313**(5791): 1304–1307. doi:10.1126/science.1127980. PMID 16946073.

63. Pollard, KS, Salama, SR, Lambert, N, Lambot, MA, Coppens, S, Pedersen, JS, Katzman, S, King, B, Onodera, C, Siepel, A, Kern, AD, Dehay, C, Igel, H, Ares, M Jr, Vanderhaeghen, P, Haussler, D (2006). An RNA gene expressed during cortical development evolved rapidly in humans. *Nature*, **443**(7108): 167–172. doi:10.1038/nature05113. PMID 16915236 (supplement).

64. https://www.nobelprize.org/nobel_prizes/medicine/laureates/1908/mechnikov-facts.html

65. Claesson, MJ, et al. (2012). Gut microbiota composition correlates with diet and health in the elderly. *Nature*, **488**(7410): 178–184.

66. Woodmansey, EJ (2007). Intestinal bacteria and ageing. *J Appl Microbiol*, **102**(5): 1178–1186.

67. O'Toole, W, Claesson, MJ (2010). Gut microbiota: changes throughout the lifespan from infancy to elderly. *Int Dairy J*, **20**: 281–291.

68. https://www.nobelprize.org/nobel_prizes/medicine/laureates/1958/lederberg-facts.html

69. Bäckhed, F, et al. (2005). Host-bacterial mutualism in the human intestine. *Science*, **307**(5717): 1915–1920.

70. Sivan, A, Corrales, L, et al. (2015). Commensal Bifidobacterium promotes antitumor immunity and facilitates anti–PD-L1 efficacy. *Science*, **350**(6264): 1084–1089.

71. Ahn, J, et al. (2013). Human gut microbiome and risk for colorectal cancer. *J Natl Cancer Inst*, **105**(24): 1907–1911.

72. Zackular, JP, et al. (2014). The human gut microbiome as a screening tool for colorectal cancer. *Cancer Prev Res (Phila)*, **7**(11): 1112–1121.

73. El-Abbadi, NH, Dao, MC, Meydani, SN (2014). Yogurt: role in healthy and active aging. *Am J Clin Nutr*, **99**(5 Suppl): 1263S–1270S. doi: 10.3945/ajcn.113.073957. http://www.ncbi.nlm.nih.gov/pubmed/24695886

74. Vétizou, M, et al. (2015). Anticancer immunotherapy by CTLA-4 blockade relies on the gut microbiota. *Science*, **350**(6264): 1079–1084.

75. Shreiner, AB, Kao, JY, Young, VB (2015). The gut microbiome in health and in disease. *Curr Opin Gastroenterol*, **31**: 69–75.

76. Wesemann, DR, et al. (2013). Microbial colonization influences early B-lineage development in the gut lamina propria. *Nature*, **501**(7465): 112–115.

77. Arumugam, M, et al. (2011). Enterotypes of the human gut microbiome. *Nature*, **473**(7346): 174–180.

78. Wu, GD, et al. (2011). Linking long-term dietary patterns with gut microbial enterotypes. *Science*, **334**(6052): 105–108.

79. Qin, J, et al. (2012). A metagenome-wide association study of gut microbiota in type 2 diabetes. *Nature*, **490**(7418): 55–60.

80. Atarashi, K, et al. (2011). Induction of colonic regulatory T cells by indigenous Clostridium species. *Science*, **331**(6015): 337–341.

81. Bollinger, R, et al. (2007). Biofilms in the Large Bowel suggest an apparent function of the Human Vermiform Appendix. *J Theor Biol*, **249**(4): 826–831.

82. https://www.sciencemag.org/news/2013/02/appendix-evolved-more-30-times

83. Rankin LC, et al. (2016). Complementarity and redundancy of IL-22-producing innate lymphoid cells. *Nat Immunol*, **17**(2): 179–186.

84. Zahid, A. (2004). The vermiform appendix: not a useless organ. *J Coll Physicians Surg Pak*, **14**(4): 256–258.

85. http://dukemagazine.duke.edu/article/a-little-dirt-wont-hurt

86. Killinger, BA, et al. (2018). The vermiform appendix impacts the risk of developing Parkinson's disease. *Sci Transl Med*, **10**(465): pii:eaar5280.

87. Hsiao, EY, et al. (2013). Microbiota modulate behavioral and physiological abnormalities associated with neurodevelopmental disorders. *Cell*, **155**(7): 1451–1463.

88. Shen, HH (2015). New feature: microbes on the mind. *Proc Natl Acad Sci U S A*, **112**(30): 9143–9145.

89. Hsiao, EY, et al. (2013). Microbiota modulate behavioral and physiological abnormalities associated with neurodevelopmental disorders. *Cell*, **155**(7): 1451–1463.

90. Turnbaugh, PJ, et al. (2009). A core gut microbiome in obese and lean twins. *Nature*, **457**(7228): 480–484.

91. Greenblum, S, et al. (2012). Metagenomic systems biology of the human gut microbiome reveals topological shifts associated with obesity and inflammatory bowel disease. *Proc Natl Acad Sci U S A*, **109**(2): 594–599. https://www.hopkinsmedicine.org/news/media/releases/blood_vessels_sniff_gut_microbes_to_regulate_blood_pressure

92. Xavier, RJ, Podolsky, DK (2007). Unravelling the pathogenesis of inflammatory bowel disease. *Nature*, **448**(7152): 427–434.

93. Ridaura, VK, et al. (2013). Gut microbiota from twins discordant for obesity modulate metabolism in mice. *Science*, **341**(6150): 1241214.

94. Smith, PA (2015). Brain meet gut neuroscientists are probing the connections between intestinal microbes and brain development. *Nature*, **526**: 312–314.

95. Greenblum, S, et al. (2012). Metagenomic systems biology of the human gut microbiome reveals topological shifts associated with obesity and inflammatory bowel disease. *Proc Natl Acad Sci U S A*, **109**(2): 594–599.

96. Diaz Heijtz, R, et al. (2011). Normal gut microbiota modulates brain development and behavior. *Proc Natl Acad Sci U S A*, **108**(7): 3047–3052.

97. Mayer, EA, et al. (2014). Altered brain-gut axis in autism: comorbidity or causative mechanisms? *Bioessays*, **36**(10): 933–939.

98. http://www.sciencemag.org/news/2018/06/fecal-transplants-might-help-save-vulnerable-koalas?utm_campaign=news_daily_2018-06-18&et_rid=34842998&et_cid=2122885

99. Tilg, H, Moschen, AR (2014). Microbiota and diabetes: an evolving relationship. *Gut*, **63**(9): 1513–1521.

100. Musso, G, et al. (2010). Obesity, diabetes, and gut microbiota: the hygiene hypothesis expanded? *Diabetes Care*, 3(10): 2277–2284.

101. http://www.metahit.eu/

102. https://www.ncbi.nlm.nih.gov/pubmed/19819907

103. http://www.earthmicrobiome.org/

104. Ashburner, M, Ball, CA, Blake, JA, Botstein, D, Butler, H, Cherry, JM, et al. (2000). Gene ontology: tool for the unification of biology. The gene ontology consortium. *Nat Genet*, **25**(1): 25–29.

105. Salton, MR, Kim, KS (1996). Structure. In Baron, S, (ed.) *Medical Microbiology*, 4th ed. Univ of Texas Medical Branch, Galveston. ISBN 0-9631172-1-1.

106. Stefanini, I, Dapporto, L, Legras, J-L, Calabretta, A, Di Paola, M, De Filippo, C, Viola, R, Capretti, P, Polsinelli, M, Turillazzi, S, Cavalieri, D (2012). Saccharomyces cerevisiae and social wasps. *Proc Natl Acad Sci*, **109**(33): 13398–13403; doi:10.1073/pnas.1208362109.

107. https://www.nobelprize.org/prizes/medicine/2002/summary/

108. https://www.nobelprize.org/prizes/medicine/2006/summary/

109. https://www.nobelprize.org/prizes/chemistry/2008/summary/

110. https://www.nobelprize.org/prizes/medicine/1933/morgan/facts/

111. http://science.sciencemag.org/content/66/1699/84/tab-e-letters

112. https://www.nobelprize.org/prizes/medicine/1946/muller/lecture/

113. https://www.nobelprize.org/prizes/medicine/1958/summary/

114. https://www.nobelprize.org/prizes/medicine/1995/7713-the-nobel-prize-in-physiology-or-medicine-1995/

115. https://www.nobelprize.org/prizes/medicine/2004/press-release/

116. https://www.nobelprize.org/prizes/medicine/2009/summary/

117. https://www.nobelprize.org/prizes/medicine/2011/press-release/

118. https://www.nobelprize.org/prizes/medicine/2017/summary/

119. Zhou, D, Xue, J, Lai, JCK, Schork, NJ, White, KP, Haddad, GG (2008). Mechanisms underlying hypoxia tolerance in drosophila melanogaster: hairy as a metabolic switch. *PLoS Genet*, https://doi.org/10.1371/journal.pgen.1000221

120. https://www.jax.org/

121. https://www.nobelprize.org/nobel_prizes/medicine/laureates/1908/ehrlich-facts.html

122. https://www.nobelprize.org/nobel_prizes/medicine/laureates/1984/

123. http://www.nobelprize.org/nobel_prizes/medicine/laureates/2007/press.html

124. Settembre, C, et al. (2007). Systemic inflammation and neurodegeneration in a mouse model of multiple sulfatase deficiency. *Proc Natl Acad Sci U S A*, **104**(11): 4506–4511.

125. The African clawed frog, much used in embryological research. It can also be used in human pregnancy testing, as it produces eggs in response to substances in the urine of a pregnant woman.

126. Chinwalla, AT, et al. (2002). Initial sequencing and comparative analysis of the mouse genome. *Nature*, **420**: 520–562.

127. Transcription activator-like effector nucleases (TALEN) are restriction enzymes that can be engineered to cut specific sequences of DNA.

128. Pan, B, Akyuz, N, Liu, X-P, Asai, Y, Nist-Lund, C, Kurima, K, Derfler, BH, György, B, Limapichat, W, Walujkar, S, Wimalasena, LN, Sotomayor, M, Corey, DP, Holt, JR (2018). TMC1 forms the pore of mechanosensory transduction channels in vertebrate inner ear hair cells. *Neuron*, **99**, 736–753.

129. https://news.harvard.edu/gazette/story/2018/08/hearing-protein/

130. Askew, C, et al. (2015). TMC gene therapy restores auditory function in deaf mice. *Sci Transl Med*, **7**(295): 295ra108.

131. https://www.nobelprize.org/nobel_prizes/medicine/laureates/2000/kandel-lecture.html

132. Kandel, ER (2006). *In Search of Memory*. W.W. Norton & Company, New York.

133. https://www.nobelprize.org/nobel_prizes/medicine/laureates/1973/press.html

134. https://www.nobelprize.org/nobel_prizes/medicine/laureates/2000/presentation-speech.html

135. Kandel, ER, Nobel Lecture. https://www.nobelprize.org/nobel_prizes/medicine/laureates/2000/kandel-lecture.html

136. Kandel, ER (2006). *In Search of Memory*, p. 236.

137. https://www.amazon.com/Genetics-Species-Columbia-Classics-Evolution/dp/0231054750/ref=sr_1_3?ie=UTF8&qid=1535492490&sr=8-3&keywords=genetics+and+the+origin+of+species&dpID=514HIZWKQxL&preST=_SY344_BO1,204,203,200_QL70_&dpSrc=srch

138. http://thenode.biologists.com/doctor-delayed-publications-remarkable-life-george-streisinger/careers/

139. Streisinger, G, Walker, C, Dower, N, Knauber, D, Singer, F (1981). Production of clones of homozygous diploid zebra fish (Brachydanio rerio). *Nature*, **291**: 293–296.

140. https://zfin.org/

141. https://zebrafish.org/home/guide.php

142. Mills, D (1993). *Eyewitness Handbook: Aquarium Fish*. Harper Collins. ISBN 0-7322-5012-9.

143. http://en.zfish.cn/

144. Lamason, RL, Mohideen, MA, Mest, JR, Wong, AC, Norton, HL, Aros, MC, Jurynec, MJ, Mao, X, Humphreville, VR, Humbert, JE, Sinha, S, Moore, JL, Jagadeeswaran, P, Zhao, W, Ning, G, Makalowska, I, McKeigue, PM, O'Donnell, D, Kittles, R, Parra, EJ, Mangini, NJ, Grunwald, DJ, Shriver, MD, Canfield, VA, Cheng, KC (2005). SLC24A5, a putative cation exchanger, affects pigmentation in zebrafish and humans. *Science*, **310**(5755): 1782–1786.

145. White, RM, Sessa, A, Burke, C, Bowman, T, LeBlanc, J, Ceol, C, Bourque, C, Dovey, M, Goessling, W, Burns, CE, Zon, LI (2008). Transparent adult zebrafish as a tool for in vivo transplantation analysis. *Cell Stem Cell*, **2**(2): 183–189.

146. https://www.sanger.ac.uk/science/programmes/archive-mouse-and-zebrafish-genetics-programme

147. https://zfin.org/ZDB-LAB-971209-4

148. https://www.ncbi.nlm.nih.gov/grc

149. http://archive.indianexpress.com/news/decoding-the-genome-mystery/485122

150. Howe, K, Clark, MD, Torroja, CF, Torrance, J, Berthelot, C, Muffato, M, et al. (2013). The zebrafish reference genome sequence and its relationship to the human genome. *Nature*, **496**(7446): 498–503.

151. https://irp.nih.gov/blog/post/2016/08/why-use-zebrafish-to-study-human-diseases

152. Dahm, R (2006). The zebrafish exposed. *Am Sci*, **94**(5): 446–453.

153. Goldshmit, Y, Sztal, TE, Jusuf, PR, Hall, TE, Nguyen-Chi, M, Currie, PD (2012). Fgf-dependent glial cell bridges facilitate spinal cord regeneration in zebrafish. *J Neurosci*, **32**(22): 7477–7492.

154. Head, JR, Gacioch, L, Pennisi, M, Meyers, JR (2013). Activation of canonical Wnt/β-catenin signaling stimulates proliferation in neuromasts in the zebrafish posterior lateral line. *Dev Dyn*, **242**(7): 832–846.

155. Bassett, DI, Currie, PD (2003). The zebrafish as a model for muscular dystrophy and congenital myopathy. *Hum Mol Genet*, **12**(Spec No 2): R26570.

156. Wilbanks, AM, Fralish, GB, Kirby, ML, Barak, LS, Li, Y-X, Caron, MG (2004). ß-arrestin 2 regulates zebrafish development through the hedgehog signaling pathway. *Science*, **306**(5705): 2264–2267.

157. Ceol, CJ, Houvras, Y, Jane-Valbuena, J, Bilodeau, S, Orlando, DA, Battisti, V, Fritsch, L, Lin, WM, Hollmann, TJ, Ferré, F, Bourque, C, Burke, CJ, Turner, L, Uong, A, Johnson, LA, Beroukhim, R, Mermel, CH, Loda, M, Ait-Si-Ali, S, Garraway, LA, Young, RA, Zon, LI (2011). The histone methyltransferase SETDB1 is recurrently amplified in melanoma and accelerates its onset. *Nature*, **471**(7339): 513–517.

158. Arthritis drug could help beat melanoma skin cancer, study finds. *Science Daily*. March 24, 2011. https://www.sciencedaily.com/releases/2011/03/110323141838.htm

159. Drummond, IA (2005). Kidney development and disease in the zebrafish. *J Am Soc Nephrol*, **16**(2): 299–304.

160. Investigating inflammatory disease using zebrafish. *Fish for Science*. http://www.fishforscience.org/disease/inflammatory-disease/

161. Novoa, B, Figueras, A (2012). Lambris, JD, Hajishengallis, G, (eds.) *Current Topics in Innate Immunity II*. Advances in Experimental Medicine and Biology. Springer New York, pp. 253–275.

162. Ramakrishnan, L (2013). Looking within the zebrafish to understand the tuberculous granuloma. *Adv Exp Med Biol*, **783**: 251–266.

163. Allison, WT, Barthel, LK, Skebo, KM, Takechi, M, Kawamura, S, Raymond, PA (2010). Ontogeny of cone photoreceptor mosaics in zebrafish. *J Comp Neurol*, **518**(20): 4182–4195.

164. Lawrence, JM, Singhal, S, Bhatia, B, Keegan, DJ, Reh, TA, Luthert, PJ, Khaw, PT, Limb, GA (2007). MIO-M1 cells and similar muller glial cell lines derived from adult human retina exhibit neural stem cell characteristics. *Stem Cells*, **25**(8): 2033–2043.

165. Plantié, E, Migocka-Patrzałek, M, Daczewska, M, Jagla, K (2015). Model organisms in the fight against muscular dystrophy: lessons from drosophila and Zebrafish. *Molecules*, **20**(4): 6237–6253.

166. Jones, KJ, Morgan, G, Johnston, H, Tobias, V, Ouvrier, RA, Wilkinson, I, North, KN (2001). The expanding phenotype of laminin alpha2 chain (merosin) abnormalities: case series and review. *J Med Genet*, **38**(10): 649–657.

167. http://science.sciencemag.org/content/361/6405/888.14

168. Fry, BG, Casewell, NR, Wüster, W, Vidal, N, Young, B, Jackson, TN (2012). The structural and functional diversification of the Toxicofera reptile venom system. *Toxicon*, **60**(4): 434–448.

169. King, GF (2011). Venoms as a platform for human drugs: translating toxins into therapeutics. *Expert Opin Biol Ther*, **11**: 1469–1484, doi:10.1517/14712598.2011.621940.

170. https://www.ncbi.nlm.nih.gov/pmc/articles/PMC5725072/

171. Li, B, Lyu, P, Xi, X, Ge, L, Mahadevappa, R, Shaw, C, Kwok, HF (2018) Triggering of cancer cell cycle arrest by a novel scorpion venom-derived peptide—gonearrestide. *J Cell Mol Med*, **22**(9): 4460–4473.

172. Schudel, M (2011-06-18). Bill Haast dies at 100: Florida snake man provided venom for snakebite serum. *The Washington Post*.

173. Jacobs, J (1890). *English Fairy Tales*. Oxford University, p. 69.

174. Gerstein, HC, Waltman, L (2006). Why don't pigs get diabetes? Explanations for variations in diabetes susceptibility in human populations living in a diabetogenic environment. *CMAJ*, **174**(1): 25–26.

175. Whitfield, J (2003). Fat pigs ape obese humans. www.nature.com>news 030804-5.

176. https://www.nobelprize.org/prizes/medicine/1923/banting/lecture/

177. Luo, Y, Lin, L, Bolund, L, Jensen, TG, Sørensen, CB (2012). Genetically modified pigs for biomedical research. *J Inherit Metab Dis*, **35**(4): 695–713.

178. Kitamura, M, et al. (2014). Glottic regeneration with a tissue-engineering technique, using acellular extracellular matrix scaffold in a canine model. *J Tissue Eng Regen Med*, **10**(10): 825–832.

179. Sicari, BM, Rubin, JP, Dearth, CL, Wolf, MT, Ambrosio, F, Boninger, M, Turner, NJ, Weber, DJ, Simpson, TW, Wyse, A, Brown, EHP, Dziki, JL, Fisher, LE, Brown, S, Badylak, SF (2014). An acellular biologic scaffold promotes skeletal muscle formation in mice and humans with volumetric muscle loss. *Sci Transl Med*, **6**(234): 234ra58.

180. https://www.foxnews.com/health/why-pigs-are-so-valuable-for-medical-research

181. https://www.nih.gov/news-events/nih-research-matters/long-lived-pig-primate-heart-transplants

182. https://www.nature.com/articles/ncomms11138

183. https://www.houstoniamag.com/articles/2017/1/23/texas-heart-institute-new-research-ghost-heart

184. https://www.egenesisbio.com/wp-content/uploads/eGenesis_Science_Aug17_final.pdf

185. https://www.onegreenplanet.org/animalsandnature/phenomenal-reasons-to-love-pigs/

186. https://www.wildrepublic.com/en/facts-about-pigs

187. https://www.cites.org/. The aim of CITES is to ensure that international trade in specimens of wild animals and plants does not threaten their survival.

188. https://onekindplanet.org/animal/elephant/

189. https://www.nature.com/articles/d41586-018-06820-4

190. http://maryvogas.com/blog/what-can-we-learn-from-elephants

191. https://www.sciencedaily.com/releases/2018/02/180226152725.htm

192. https://www.genomeweb.com/sequencing/sequencing-extinct-and-living-elephants-reveals-genetic-admixture#.W-KObqe74Us

193. Speakman, JR (2005). Body size, energy metabolism and lifespan. *J Exp Biol*, **208**(Pt 9): 1717–1730.

194. Sulak, et al. (2016). *eLife*, **5**: e11994.

195. Peto, R, Professor of Medical Statistics and Epidemiology at the University of Oxford, England.

196. Gewin, V (2013). Massive animals may hold secrets of cancer suppression. *Nature News*, 10.1038/nature.2013.12258.

197. Abegglen, LM, et al. (2015). *JAMA*, **314**(17): 1850–1860. doi:10.1001/jama.2015.13134.

198. http://www.sciencemag.org/news/2018/08/elephants-rarely-get-cancer-thanks-zombie-gene?utm_campaign=news_daily_2018-08-27&et_rid=34842998&et_cid=2323355

199. https://www.cell.com/cell-reports/fulltext/S2211-1247(18)31145-8

200. Vazquez, JM, et al. (2018). A zombie LIF gene in elephants is upregulated by TP53 to induce apoptosis in response to DNA damage. *Cell Rep*, **24**(7): 1765–1776.

201. https://www.haaretz.com/science-and-health/.premium-elephants-help-researchers-make-stride-toward-cancer-cure-1.5407873

202. https://www.theguardian.com/science/2013/jul/14/naked-mole-rat-cancer-research

203. https://www.wired.co.uk/article/naked-mole-rats-amazing

204. http://www.pnas.org/content/early/2018/08/21/1720530115.short?rss=1

205. http://www.sciencemag.org/news/2018/08/eating-poop-makes-naked-mole-rats-more-motherly

206. http://www.sciencemag.org/news/2018/01/naked-mole-rats-defy-biological-law-aging

207. Park, TJ, et al. (2017). Fructose-driven glycolysis supports anoxia resistance in the naked mole-rat. *Science*, **356**(6335): 307–311. doi: 10.1126/science.aab3896

208. Omerbašić, D, et al. (2016). Hypofunctional TrkA accounts for the absence of pain sensitization in the African naked mole-rat. *Cell Rep*, **17**(3): 748–758.

209. Seluanov, A, Hine, C, Azpurua, J, Feigenson, M, Bozzella, M, Mao, Z, Catania, KC, Gorbunova, V (2009). Hypersensitivity to contact inhibition provides a clue to cancer resistance of naked mole-rat. *Proc Natl Acad Sci U S A*. **106**(46): 19352–19357.

210. Tian, X, Azpurua, J, Hine, C, Vaidya, A, Myakishev-Rempel, M, Ablaeva, J, Mao, Z, Nevo, E, Gorbunova, V, Seluanov, A (2013). High-molecular-mass hyaluronan mediates the cancer resistance of the naked mole rat. *Nature*, **499**(7458): 346–349.

211. Azpurua, J, Ke, Z, Chen, IX, Zhang, Q, Ermolenko, DN, Zhang, ZD, Gorbunova, V, Seluanov, A (2013). Naked mole-rat has increased translational fidelity compared with the mouse, as well as a unique

28S ribosomal RNA cleavage. *Proc Natl Acad Sci U S A*, **110**(43): 17350–17355.

212. Gompertz, B (1825). On the nature of the function expressive of the law of human mortality, and on a new mode of determining the value of life contingencies. *Philos Trans R Soc*, **115**: 513–585.

213. https://www.iflscience.com/plants-and-animals/naked-molerats-break-the-rules-of-aging/

214. https://www.nobelprize.org/prizes/chemistry/2008/tsien/autobiography/

215. http://www.quincybioscience.com/products-prevagen/

216. http://science.sciencemag.org/content/342/6164/1242592

217. https://www.nature.com/articles/nature13400

218. https://news.nationalgeographic.com/news/2013/12/131212-comb-jelly-ctenophore-ocean-animals-evolution-science/

219. Olivera, BM, Teichert, RW (2007). Diversity of the neurotoxic Conus peptides: a model for concerted pharmacological discovery. *Mol Interventions*, **7**(5): 251–260.

220. https://www.nature.com/articles/nsmb.3292

221. Ramasamy, MS, Manikandan, S (2011). Novel pharmacological targets from Indian cone snails. *Mini Rev Med Chem*, **11**(2): 125–130.

222. https://www.accessdata.fda.gov/drugsatfda_docs/nda/2004/21-060_Prialt.cfm

223. www.pharmatimes.com/news/astrazeneca_signs_deal_for_pain_drugs_from_sea_snails_981596

224. Bishop, BM, Juba, ML, Russo, PS, Devine, M, Barksdale, SM, Scott, S, Settlage, R, Michalak, P, Gupta, K, Vliet, K, Schnur, JM, van Hoek, ML (2017). Discovery of novel antimicrobial peptides from varanus komodoensis (komodo dragon) by large-scale analyses and de-novo-assisted sequencing using electron-transfer dissociation mass spectrometry. *J Proteome Res*, **16**(4), 1470–1482.

225. Chung, EMC, Dean SN, Propst CN, Bishop BM, van Hoek ML (2017). Komodo dragon-inspired synthetic peptide DRGN-1 promotes wound-healing of a mixed-biofilm infected wound. *NPJ Biofilms Microbiomes*, **3**, Article no. 9.

226. Maruno, K, Said, SI (1993). Small-cell lung carcinoma: inhibition of proliferation by vasoactive intestinal peptide and helodermin and enhancement of inhibition by anti-bombesin antibody. *Life Sci*, **52**(24): PL267– PL271.

227. https://www.sciencedaily.com/releases/2007/07/070709175815.htm

228. https://www.cardiovascularbusiness.com/topics/prevention/ada-byetta-trumps-lantus-glucose-control-diabetics

229. https://www.livescience.com/34513-how-salamanders-regenerate-lost-limbs.html

230. Kumar, A, Godwin, JW, Gates, PB, Garza-Garcia, AA, Brockes, JP (2007). Molecular basis for the nerve dependence of limb regeneration in an adult vertebrate. *Science*, **318**(5851), 772–777.

231. Sun, C, Shepard, DB, Chong, RA, López Arriaza, J, Hall, K, Castoe, TA, Feschotte, C, Pollock, DD, Mueller, RL (2012). LTR retrotransposons contribute to genomic gigantism in plethodontid salamanders. *Genome Biol Evol*, **4**(2): 168–183. doi:10.1093/gbe/evr139.

   Elewa, A, Wang, H, Talavera-López, C, Joven, A, Brito, G, Kumar, A, Hameed, LS, Penrad-Mobayed, M, Yao, Z (2017). Reading and editing the Pleurodeles waltl genome reveals novel features of tetrapod regeneration. *Nat Commun*, **8**(1). doi:10.1038/s41467-017-01964-9. ISSN 2041-1723.

232. Nowoshilow, S, et al. (2018). The axolotl genome and the evolution of key tissue formation regulators. *Nature*, **554**(7690): 50–55. doi:10.1038/nature25458. ISSN 1476-4687.

233. https://www.nytimes.com/2017/05/17/magazine/when-the-lab-rat-is-a-snake.html

234. Adams, G (2012-02-01). Pythons are squeezing the life out of the Everglades, scientists warn. *The Independent*, London.

235. https://www.ua.edu/news/2016/10/exploring-possible-link-between-pythons-diabetes-in-humans/

236. https://www.colorado.edu/biofrontiers/2011/12/12/leslie-leinwand-discusses-python-project

237. https://www.sciencedaily.com/releases/2017/06/170622121906.htm

238. https://www.ncbi.nlm.nih.gov/pmc/articles/PMC3870669/

239. Palmer, BA, Taylor, GJ, Brumfeld, V, Gur, D, Shemesh, M, Elad, N, Osherov, A, Oron, D, Weiner, S, Addadi, L (2017). The image-forming mirror in the eye of the scallop. *Science*, **358**: 1172–1175.

240. https://www.nobelprize.org/nobel_prizes/chemistry/laureates/2017/

241. http://www.pbs.org/wnet/nature/owl-power/11628/

242. http://www.hopkinsmedicine.org/news/

243. http://www.bostonretinalimplant.org/

244. Brindley, G, Lewin, W (1968). The sensation produced by electrical stimulation of the visual cortex. *J Physiol*, **196**: 479–493.

245. Zrenner, E (2002). Will retinal implants restore vision? *Science*, **295**: 1022–1025.

246. https://journals.plos.org/plosone/article?id=10.1371/journal.pone.0115239

247. http://www.secondsight.com/g-the-argus-ii-prosthesis-system-pf-en.html

248. https://www.retina-implant.de/en/implant/ri-alpha-ams/

249. http://www.pixium-vision.com/en/technology-1/iris-vision-restoration-system

250. Bakken, GS, Krochmal, AR (2007). The imaging properties and sensitivity of the facial pits of pitvipers as determined by optical and heat-transfer analysis. *J Exp Biol*, **210**(16): 2801–2810.

251. Neuweiler, G (1984). Foraging, echolocation and audition in bats. *Naturwissenschaften*, **71**(9): 446–455.

252. https://asa.scitation.org/doi/10.1121/1.5067725

253. http://advances.sciencemag.org/content/4/7/eaar7428

254. Johansen, K, Lenfant, C, Schmidt-Nielsen, K, et al. (1968). Gas exchange and control of breathing in the electric eel, electrophorus electricus. *Z. Vergl. Physiologie*, **61**: 137–163.

255. Foucault, et al. (1987). Quantifying aspects of shock discharge by the electric eel. *Nature*, **109**(45): 45–60.

256. https://triz-journal.com/biology-electric-eel/

257. Catania, KC (2015). Electric eels concentrate their electric field to induce involuntary fatigue in struggling prey. *Curr Biol*, **25**(22): 1–10.

258. Xu, J, Lavan, D (2008). Designing artificial cells to harness the biological ion concentration gradient. *Nat Nanotechnol*, **3**(11): 666–670.

259. http://www.sciencemag.org/news/2018/03/how-ghost-knifefish-became-fastest-electrical-discharger-animal-kingdom

260. Thompson, A, Infield, DT, Smith, AR, Smith, GT, Ahern, CA, Zakon, HH (2018). Rapid evolution of a voltage-gated sodium channel gene in a lineage of electric fish leads to a persistent sodium current. *PLoS Biol*, https://doi.org/10.1371/journal.pbio.2004892

261. Au, Whitlow WL (2012). *The Sonar of Dolphins*. Springer Science & Business Media.

262. Watson, KK, Jones, TK, Allman, JM (2006). Dendritic architecture of the Von Economo neurons. *Neuroscience*, **141**(3): 1107–1112.

263. Johnson, SP, Catania, JM, Harman, RJ, Jensen, ED (2012). Adipose-derived stem cell collection and characterization in bottlenose dolphins (Tursiops truncatus). *Stem Cells Dev*, **21**(16): 2949–2957.

264. https://www.chapala.com/chapala/magnifecentmexico/hummingbirds/hummingbirds.html

265. Lisney, TJ, Wylie, DR, Kolominsky, J, Iwaniuk, AN (2015). Eye morphology and retinal topography in hummingbirds (Trochilidae: Aves). *Brain Behav Evol*, **86**(3–4): 176–190.

266. http://rsbl.royalsocietypublishing.org/content/early/2012/02/14/rsbl.2011.1180

267. https://www.sciencenews.org/blog/wild-things/tiny-hummingbirds-can-fly-long-long-way

268. https://www.chapala.com/chapala/magnifecentmexico/hummingbirds/hummingbirds.html

269. https://defenders.org/hummingbirds/basic-facts

270. Harpole, T (1 March 2005). Falling with the Falcon. Smithsonian Air & Space magazine. Retrieved 4 September 2008.

271. https://prezi.com/fepzvrygskvs/the-respiratory-system-of-birds-of-prey/

272. Hawkes, LA, Balachandran, S, Batbayar, N, Butler, PJ, Frappell, PB, Milsom, WK, Tseveenmyadag, N, Newman, SH, Scott, GR (2011). The trans-Himalayan flights of bar-headed geese (Anser indicus). *Proc Natl Acad Sci U S A*, **108**(23): 9516–9519.

273. Swan, LW (1970). Goose of the Himalayas. *Nat Hist*, **70**: 68–75.

274. Scott, GR, Hawkes, LA, Frappell, PB, Butler, PJ, Bishop, CM, Milsom, WK (2015). How bar-headed geese fly over the Himalayas. *Physiology (Bethesda)*, **30**(2): 107–115.

275. Bishop, CM, Spivey, RJ, Hawkes, LA, Batbayar, N, Chua, B, Frappell, PB, Milsom, WK, Natsagdorj, T, Newman, SH, Scott, GR, Takekawa, JY, Wikelski, M, Butler, PJ (2015). The roller coaster flight strategy of bar-headed geese conserves energy during Himalayan migrations. *Science*, **347**(6219): 250–254.

276. West, JB (2009). Comparative physiology of the pulmonary blood-gas barrier: the unique avian solution. *Am J Physiol Regul Integr Comp Physiol*, **297**(6): R1625– R1634.

277. http://yampayaks.com/2017/06/25/yak-description-habitat-image-diet-and-fascinating/

278. Wiener, G, Jianlin, H, Ruijun, L (2003). The yak in relation to its environment, in *The Yak*, 2nd ed. Regional Office for Asia and the Pacific Food and Agriculture Organization of the United Nations, Bangkok, Chapter 4. ISBN 92-5-104965-3.

279. Sarkar, M, Das, DN, Mondal, DB (1999). Fetal haemoglobin in pregnant yaks (Poephagus grunniens L.). *Vet J*, **158**(1): 68–70.

280. https://animals.net/yak/

281. https://onekindplanet.org/animal/penguin/

282. Happy Feet is a 2006 Australian-American computer-animated musical family comedy film directed, produced, and co-written by George Miller. It stars Elijah Wood, Robin Williams, Brittany Murphy, Hugh Jackman, Nicole Kidman, Hugo Weaving, and E.G. Daily. It was produced at Sydney-based visual effects and animation studio Animal Logic for Warner Bros., Village Roadshow Pictures.

283. https://www.ncbi.nlm.nih.gov/pmc/articles/PMC4322438/

284. Li, C, et al. (2014). Two Antarctic penguin genomes reveal insights into their evolutionary history and molecular changes related to the Antarctic environment. *GigaScience*, **3**: 1–15.

285. https://www.factretriever.com/penguin-facts

286. https://www.ncbi.nlm.nih.gov/pmc/articles/PMC4089990/

287. https://www.ncbi.nlm.nih.gov/pmc/articles/PMC3077521/

288. https://www.huffpost.com/entry/10-things-we-can-learn-from-cats-that-will-make-us-happier-healthier-humans_b_8647868

289. Wang, G-D, Zhai, W, Yang, H-C, Wang, L, Zhong, L, Liu, Y-H, Fan, R-X, Yin, T-T, Zhu, C-L, Poyarkov, AD, Irwin, DM, Hytönen, MK, Lohi, H, Wu, C-I, Savolainen, P, Zhang, Y-P (2016). Out of southern East Asia: the natural history of domestic dogs across the world. *Cell Res*, **26**: 21–33. https://www.nature.com/articles/cr2015147

290. https://www.nzherald.co.nz/world/news/article.cfm?c_id=2&objectid=11157672

291. https://www.mspca.org/pet_resources/interesting-facts-about-dogs/

292. http://iditarod.com/race-map/

293. https://en.wikipedia.org/wiki/Balto_(film)

294. https://www.waltham.com/document/nutrition/dog/dog-genetics/263/

295. https://www.nationalgeographic.com.au/animals/platypus.aspx

296. https://www.nationalgeographic.com/animals/mammals/p/platypus/

297. https://www.wildrepublic.com/en/platypus

298. https://www.nature.com/scitable/topicpage/interpreting-shared-characteristics-the-platypus-genome-44568

299. https://www.nobelprize.org/nobel_prizes/chemistry/laureates/1980/berg-facts.html

300. https://www.sciencehistory.org/historical-profile/herbert-w-boyer-and-stanley-n-cohen

301. https://www.nih.gov/about-nih/what-we-do/nih-almanac/national-cancer-institute-nci

302. https://www.sciencehistory.org/historical-profile/herbert-w-boyer-and-stanley-n-cohen

303. Tswett, M (1906). Adsorption analysis and chromatographic method: application to the chemistry of chlorophyll. *Proc Ger Bot Soc*, **24**: 384–393.

304. Fiers had spent 1960–62 at the California Institute of Technology to study viral DNA, and at the University of Wisconsin in Madison to work in the laboratory of Gobind Khorana (Nobel laureate 1968 for his work on nucleic acid synthesis in the study of the genetic code).

305. Sanger, F, Nicklen, S, Coulson, AR (1977). DNA sequencing with chain-terminating inhibitors. *Proc Natl Acad Sci U S A*, **74**(12): 5463–5467.

306. Maxam, AM, Gilbert, W (1977). A new method for sequencing DNA. *Proc Natl Acad Sci U S A*, **74**(2): 560–564.

307. http://www.nobelprize.org/nobel_prizes/chemistry/laureates/1980/press.html

308. http://www.nobelprize.org/nobel_prizes/chemistry/laureates/1980/gilbert-lecture.pdf

309. http://www.nobelprize.org/nobel_prizes/medicine/laureates/1959/kornberg-lecture.pdf, page 12

310. Sanger, F died on November 19, 2013, at age 95. He won two Nobel prizes in chemistry: 1958 for protein (insulin) sequencing, 1980 for DNA sequencing.

311. Example and subsequent explanation adapted from Cook-Deegan, RM.

312. see his Nobel lecture, http://www.nobelprize.org/nobel_prizes/chemistry/laureates/1980/sanger-lecture.pdf

313. Ronaghi, M, Karamohamed, S, Pettersson, B, Uhlén, M, Nyrén, P (1996). Real-time DNA sequencing using detection of pyrophosphate release. *Anal Biochem*, **242**(1): 84–89.

314. http://en.wikipedia.org/wiki/454_Life_Sciences

315. http://en.wikipedia.org/wiki/Illumina_%28company%29

316. http://www.pacificbiosciences.com/aboutus/history/

317. Davies, K (2010). *The $1,000 Genome: The Revolution in DNA Sequencing and the New Era of Personalized Medicine*. Free Press. ISBN-10: 1416569596.

318. https://www.nanoporetech.com/

319. http://www.genome.gov/27544383

320. http://www.nobelprize.org/nobel_prizes/medicine/laureates/1908/press.html

321. http://www.nobelprize.org/nobel_prizes/medicine/laureates/1984/

322. Rascio, N, Navari-Izzo, F (2010). Heavy metal hyperaccumulating plants: How and why do they do it? And what makes them so interesting? *Plant Sci*, **180**(2): 169–181.

323. Guidi Nissim, W, Palm, E, Mancuso, S, Azzarello, E (2018). Trace element phytoextraction from contaminated soil: a case study under Mediterranean climate. *Environ Sci Pollut. Res*, https://doi.org/10.1007/s11356-018-1197-x

324. Löffler, FE, et al. (2013). Dehalococcoides mccartyi gen. nov., sp. nov., obligately organohalide-respiring anaerobic bacteria relevant to halogen cycling and bioremediation, belong to a novel bacterial class, Dehalococcoidia classis nov., order Dehalococcoidales ord. nov. and family Dehalococcoidaceae fam. nov., within the phylum Chloroflexi. *Int J Syst Evol Microbiol*, **63**(Pt 2): 625–635.

325. http://www.pnas.org/content/111/33/12103

326. Pennisi, E (2013). The CRISPR craze. *Science*, **341**(6148): 833–836.

327. Lander, ES (2016). The heroes of CRISPR. *Cell*, **164**(1–2): 18–28.

328. Ishino, Y, Shinagawa, H, Makino, K, Amemura, M, Nakata, A (1987). Nucleotide sequence of the iap gene, responsible for alkaline phosphatase isozyme conversion in Escherichia coli, and identification of the gene product. *J Bacteriol*, **169**(12): 5429–5433.

329. Hsu, PD, Lander, ES, Zhang, F (2014). Development and applications of CRISPR-Cas9 for genome engineering. *Cell*, **157**(6): 1262–1278.

330. van Soolingen, D, de Haas, PE, Hermans, PW, Groenen, PM, van Embden, JD (1993). Comparison of various repetitive DNA elements as genetic

markers for strain differentiation and epidemiology of Mycobacterium tuberculosis. *J Clin Microbiol*, **31**(8): 1987–1995.

331. Groenen, PM, Bunschoten, AE, van Soolingen, D, van Embden, JD (1993). Nature of DNA polymorphism in the direct repeat cluster of Mycobacterium tuberculosis; application for strain differentiation by a novel typing method. *Mol Microbiol*, **10**(5): 1057–1065.

332. Mojica, FJ, Montoliu, L (2016). On the origin of CRISPR-Cas technology: from prokaryotes to mammals. *Trends Microbiol*, **24**(10): 811–820.

333. Mojica, FJ, Díez-Villaseñor, C, Soria, E, Juez, G (2000). Biological significance of a family of regularly spaced repeats in the genomes of archaea, bacteria and mitochondria. *Mol Microbiol*, **36**(1): 244–246.

334. Barrangou, R, van der Oost, J (2013). *CRISPR-Cas Systems: RNA-Mediated Adaptive Immunity in Bacteria and Archaea*. Springer, Heidelberg, p. 6.

335. Mojica, FJ, Díez-Villaseñor, C, García-Martínez, J, Soria, E (2005). Intervening sequences of regularly spaced prokaryotic repeats derive from foreign genetic elements. *J Mol Evol*, **60**(2): 174–182.

336. Makarova, KS, Grishin, NV, Shabalina, SA, Wolf, YI, Koonin, EV (2006). A putative RNA-interference-based immune system in prokaryotes: computational analysis of the predicted enzymatic machinery, functional analogies with eukaryotic RNAi, and hypothetical mechanisms of action. *Biol Direct*, **1**: 7.

337. Barrangou, R, Fremaux, C, Deveau, H, Richards, M, Boyaval, P, Moineau, S, Romero, DA, Horvath, P (2007). CRISPR provides acquired resistance against viruses in prokaryotes. *Science*, **315**(5819): 1709–1712.

338. Horvath, P, Romero, DA, Coûté-Monvoisin, A-C, Richards, M, Deveau, H, Moineau, S, Boyaval, P, Fremaux, C, Barrangou, R (2008). Diversity, activity, and evolution of CRISPR loci in Streptococcus thermophilus. *J Bacteriol*, **190**(4): 1401–1412.

339. Horvath, P, Barrangou, R (2010). CRISPR/Cas, the immune system of bacteria and archaea. *Science*, **327**(5962): 167–170.

340. http://www.danisco.com/about-dupont/news/news-archive/2017/dupont-scientist-philippe-horvath-receives-franklin-institute-science-prize-2018-bower-award-for-groundbreaking-research-on-crispr-cas/

341. Pourcel, C, Salvignol, G, Vergnaud, G. (2005). CRISPR elements in Yersinia pestis acquire new repeats by preferential uptake of bacteriophage DNA, and provide additional tools for evolutionary studies. *Microbiology*, **151**(Pt 3): 653–663.

342. Bolotin, A, Quinquis, B, Sorokin, A, Ehrlich, SD (2005). Clustered regularly interspaced short palindrome repeats (CRISPRs) have spacers of extrachromosomal origin. *Microbiology*, **151**(Pt 8): 2551–2561.

343. Marraffini, LA, Sontheimer, EJ (2008). CRISPR interference limits horizontal gene transfer in staphylococci by targeting DNA. *Science*, **322**(5909): 1843–1845.

344. Deltcheva, E, Chylinski, K, Sharma, CM, Gonzales, K, Chao, Y, Pirzada, ZA, Eckert, MR, Vogel, J, Charpentier, E (2011). CRISPR RNA maturation by trans-encoded small RNA and host factor RNase II. *Nature*, **471**: 602–607.

345. Gasiunas, G, Barrangou, R, Horvath, P, Siksnys, V (2012). Cas9–crRNA ribonucleoprotein complex mediates specific DNA cleavage for adaptive immunity in bacteria. *Proc Natl Acad Sci U S A*, **109**(39): E2579–E2586.

346. Jinek, M, Chylinski, K, Fonfara, I, Hauer, M, Doudna, JA, Charpentier, E (2012). A programmable dual-RNA-guided DNA endonuclease in adaptive bacterial immunity. *Science*, **337**(6096): 816–821.

347. Horvath, P, Barangou, R (2010). CRISPR/Cas, the immune system of bacteria and archaea. *Science*, **327**(5962): 167–170.

348. https://www.broadinstitute.org/research-highlights-crispr

349. Zhang, F (2012). CRISPR-Cas systems and methods for altering expression of gene products. Patent US8697359B1.

350. Cong, L, Ran, FA, Cox, D, Lin, S, Barretto, R, Habib, N, Hsu, PD, Wu, X, Jiang, W, Marraffini, LA, Zhang, F (2013). Multiplex genome engineering using CRISPR/Cas systems. *Science*, **339**: 819–823.

351. Mali, P, Yang, L, Esvelt, KM, Aach, J, Guell, M, DiCarlo, JE, Norville, JE, Church, GM (2013). RNA-guided human genome engineering via Cas9. *Science*, **339**: 823–826.

352. Wang, H, Yang, H, Shivalila, CS, Dawlaty, MM, Cheng, AW, Zhang, F, Jaenisch, R (2013). One-step generation of mice carrying mutations in multiple genes by CRISPR/Cas-mediated genome engineering. *Cell*, **153**(4): 910–918.

353. Ledford, H (2015). CRISPR, the disruptor. *Nature*, **522**(7554): 20–24.

354. https://www.egenesisbio.com/technology/#gene-editing

355. https://www.nature.com/articles/nature23305

356. https://www.cell.com/molecular-therapy-family/molecular-therapy/fulltext/S1525-0016(18)30378-2

357. http://www.sciencemag.org/news/2018/08/scientists-tweak-dna-viable-human-embryos?utm_campaign=news_daily_2018-08-27&et_rid=34842998&et_cid=2323355

358. https://www.nature.com/articles/d41586-018-06999-6

359. https://www.nature.com/articles/d41586-018-06999-6

360. https://futurism.com/researchers-used-crispr-to-successfully-increase-hiv-resistance-in-animals/

361. https://www.technologyreview.com/s/612458/exclusive-chinese-scientists-are-creating-crispr-babies/

362. https://www.theguardian.com/science/2017/feb/16/woolly-mammoth-resurrection-scientists

363. https://www.cshl.edu/plant-geneticists-develop-new-application-crispr-break-yield-barriers-crops/

364. https://simple.wikipedia.org/wiki/Metric_system#Length

365. https://www.cnbc.com/2015/06/04/why-the-us-hasnt-fully-adopted-the-metric-system.html

366. https://en.wikipedia.org/wiki/Daniel_Gabriel_Fahrenheit

367. https://en.wikipedia.org/wiki/Kelvin

368. Samuels, DS, Radolf, JD, eds. (2010). *Borrelia: Molecular Biology, Host Interaction and Pathogenesis*. Caister Academic Press, ISBN 978-1-904455-58-5.

369. https://www.cdc.gov/onehealth/basics/zoonotic-diseases.html

370. Morens, DM, Fauci, AS (2007). The 1918 influenza pandemic: insights for the 21st century. *J Infect Dis*, **195** (7): 1018–1028.

371. La Grippe Espagnole de 1918 (in French). Institut Pasteur, internal report 2015.

372. Baxby, D (1999). Edward Jenner's inquiry; a bicentenary analysis. *Vaccine*, **17**(4): 301–307.

373. Lombard, M, Pastoret, PP, Moulin, AM (2007). A brief history of vaccines and vaccination. *Rev Sci Tech,* **26**(1): 29–48.

374. https://www.nobelprize.org/prizes/chemistry/2014/summary/

375. https://www.nobelprize.org/prizes/medicine/1975/baltimore/facts/

376. Michiels, B, Van Puyenbroeck, K, Verhoeven, V, Vermeire, E, Coenen, S (2013). Jefferson, T (ed.) The value of neuraminidase inhibitors for the prevention and treatment of seasonal influenza: a systematic review of systematic reviews. *PLoS One*, **8**(4): e60348.

377. Trifonov, V, Khiabanian, H, Rabadan, R (2009). Geographic dependence, surveillance, and origins of the 2009 influenza A (H1N1) virus. *New Engl J Med,* **361**(2): 115–119.

378. https://www.who.int/en/news-room/fact-sheets/detail/ebola-virus-disease

379. *Front Immunol,* 15 April 2014; https://doi.org/10.3389/fimmu.2014.00174

380. https://www.webmd.com/hiv-aids/guide/hiv-aids-living-managing

381. Peiris, JS, Lai, ST, Poon, LL, et al. (2003). Coronavirus as a possible cause of severe acute respiratory syndrome. *Lancet,* **361**(9366): 1319–1325.

382. Ge, X-Y, Li, J-L, Yang X-L, et al. (2013). Isolation and characterization of a bat SARS-like coronavirus that uses the ACE2 receptor. *Nature,* **503**(7477): 535–538.

383. https://www.ncbi.nlm.nih.gov/books/NBK51219/table/cerconsangina.tu1/

384. https://en.wikipedia.org/wiki/Gro_Harlem_Brundtland

385. Yardley, J (2005). After its epidemic arrival, SARS vanishes, 15 May 2005, The New York Times.

386. Zumla, A, Hui, DS, Perlman, S (2015). Middle East respiratory syndrome. *Lancet,* **386**(9997): 995–1007. https://www.ncbi.nlm.nih.gov/pmc/articles/PMC4721578/;

387. Zaki, AM, van Boheemen, S, Bestebroer, TM, Osterhaus, AD, Fouchier, RA (2012). Isolation of a novel coronavirus from a man with pneumonia in Saudi Arabia. *New Engl J Med,* **367**(19): 1814–1820.

388. Kupferschmidt, K (2020). The Coronavirus Czar. *Science,* **368**(6490), 462–464.

389. MERS situation update, January 2020. World Health Organization.

390. WHO Director-General's opening remarks at the media briefing on COVID-19—11 March 2020. World Health Organization.

391. https://www.nature.com/articles/s41586-020-2196-x_reference.pdf (received March 1, accepted March 24, 2020).

392. http://www.chinabiotoday.com/articles/vision-14-million-pathogen

393. Huang, C, Wang, Y, Li, X, Ren, L, Zhao, J, Hu, Y, et al. (2020). Clinical features of patients infected with 2019 novel coronavirus in Wuhan, China. *Lancet,* **395**(10223): 497–506.

394. https://www.gisaid.org/help/publish-with-data-from-gisaid/

395. https://www.charite.de/en/service/press_reports/artikel/detail/ neues_zentrum_fuer_globale_gesundheit_charite_global_health_ gegruendet-1/

396. https://www.cell.com/current-biology/pdf/S0960-9822(20)30662-X.pdf

397. https://en.wikipedia.org/wiki/COVID-19_testing#cite_note-sheridan-62

398. https://www.nature.com/articles/d41587-020-00002-2

399. https://diagnostics.roche.com/us/en/products/params/cobas-sars-cov-2-test.html

400. https://diagnostics.roche.com/us/en/products/params/elecsys-anti-sars-cov-2.html

401. https://www.abbott.com/corpnewsroom/product-and-innovation/ an-update-on-abbotts-work-on-COVID-19-testing.html

402. https://www.thermofisher.com/us/en/home/clinical/public-health/ coronavirus-sars-cov-2-research-solutions.html

403. Weissleder, R, Lee, H, Ko, J, Pittet, MJ (03 Jun 2020). *Science Translational Medicine*, **12**(546), eabc1931; https://stm.sciencemag. org/content/12/546/eabc1931

404. COVID-19 Diagnostic Resource Centre (2020); https://www.finddx. org/covid-19/dx-data/

405. https://www.worldometers.info/coronavirus/?utm_ campaign=homeAdUOA?Si#countries

406. https://www.cnet.com/health/how-to-get-tested-for-coronavirus-at-home/

407. Wadman, M, Couzin-Frankel, J, Kaiser, J, Matacic, C (2020). A rampage through the body. *Science*, **368**(6489), 356–360.

408. https://www.ncbi.nlm.nih.gov/pmc/articles/PMC3294426/

409. https://www.endocrineweb.com/covid-19-post-infection-thyroid-disease

410. Klok, FA, et al. (2020). Incidence of thrombotic complications in critically ill ICU patients with COVID-19; https://doi.org/10.1016/j. thromres.2020.04.013

411. https://coronavirus.health.ny.gov/childhood-inflammatory-disease-related-covid-19

412. https://www.mayoclinic.org/diseases-conditions/kawasaki-disease/ symptoms-causes/syc-20354598

413. Puelles, VG, et al. (13 May 2020). Multiorgan and renal tropism of SARS-CoV-2. *New Engl J Med*; https://www.nejm.org/doi/full/10.1056/NEJMc2011400

414. https://www.thelancet.com/journals/lancet/article/PIIS0140-6736(20)31189-2/fulltext

415. https://www.ncbi.nlm.nih.gov/pmc/articles/PMC7162753/

416. https://www.thelancet.com/journals/lancet/article/PIIS0140-6736(20)31183-1/fulltext

417. https://www.bloomberg.com/news/articles/2020-05-18/over-100-million-in-china-s-northeast-thrown-back-under-lockdown

418. https://www.worldometers.info/coronavirus/

419. https://www.who.int/emergencies/diseases/strategic-and-technical-advisory-group-for-infectious-hazards/members/biographies/en/index2.html

420. https://en.wikipedia.org/wiki/K%C3%A1ri_Stef%C3%A1nsson

421. https://www.niaid.nih.gov/about/director

422. Cohen, J (13 April 2020). Mice, hamsters, ferrets, monkeys. Which animals can help defeat the new coronavirus? *Sci Mag* ; https://pulitzercenter.org/reporting/mice-hamsters-ferrets-monkeys-which-lab-animals-can-help-defeat-new-coronavirus

423. https://science.sciencemag.org/content/early/2020/05/06/science.abc1932

424. https://www.biorxiv.org/content/10.1101/2020.03.17.995639v1

425. https://www.biorxiv.org/content/10.1101/2020.03.13.990226v1

426. https://jvi.asm.org/content/81/2/813

427. https://www.nature.com/articles/s41564-018-0317-1

428. https://science.sciencemag.org/content/early/2020/04/07/science.abb7015

429. https://www.nytimes.com/interactive/2020/science/coronavirus-vaccine-tracker.html

430. https://www.nature.com/articles/d41586-020-01221-y

431. https://www.reuters.com/article/us-health-coronavirus-moderna-funding/moderna-receives-483-million-barda-award-for-covid-19-vaccine-development-idUSKBN21Y3E0

432. https://globalbiodefense.com/newswire/barda-and-sanofi-prepare-for-studies-of-covid-19-vaccine/

433. https://www.fiercepharma.com/pharma/astrazeneca-scores-1b-from-u-s-signs-up-to-deliver-hundreds-millions-covid-19-vaccines

434. Sheridan, C (May 2020). Convalescent serum ...; https://www.nature.com/articles/d41587-020-00011-1

435. https://www.mayoclinic.org/drugs-supplements/hydroxychloroquine-oral-route/side-effects/drg-20064216

436. https://www.fda.gov/media/137565/download

437. https://www.mayoclinic.org/drugs-supplements/dexamethasone-oral-route/description/drg-20075207

438. https://www.covig-19plasmaalliance.org/en-us#recruitment

439. https://www.takeda.com/en-us/newsroom/covid-19-updates/

440. https://www.cslbehring.com/newsroom/2020/covid-19-update

441. https://www.grifols.com/en/bioscience

442. http://www.bplgroup.com/about-bpl-group/covid-19-updates/

443. https://www.groupe-lfb.com/en/products-and-activities/medicinal-products/immunology/

444. https://www.octapharma.com/about-us/who-we-are/

445. https://www.fiercebiotech.com/biotech/gigagen-jumps-into-covid-19-arena-polyclonal-antibodies

446. https://hub.jhu.edu/2020/04/08/arturo-casadevall-blood-sera-profile/

447. https://www.gilead.com/purpose/advancing-global-health/covid-19/remdesivir-clinical-trials

448. https://www.nejm.org/doi/full/10.1056/NEJMoa2007764

449. https://www.sciencemag.org/news/2020/06/cheap-steroid-first-drug-shown-reduce-death-covid-19-patients?utm_campaign=news_daily_2020-06-16&et_rid=34842998&et_cid=3368179

450. https://newsroom.regeneron.com/news-releases/news-release-details/regeneron-announces-expanded-collaboration-hhs-develop-antibody

451. https://petrov.stanford.edu/pdfs/0124.pdf

452. https://royalsocietypublishing.org/doi/10.1098/rspb.2016.2094

453. https://www.yesmagazine.org/issues/coronavirus-community-power/

Index